NUMERICAL MATHEMATICS AND SCIENTIFIC COMPUTATION

Series Editors

G. H. GOLUB Ch. SCHWAB
E. SÜLI

NUMERICAL MATHEMATICS AND SCIENTIFIC COMPUTATION

Books in the series

Monographs marked with an asterix (*) appeared in the series 'Monographs in Numerical Analysis' which has been folded into, and is continued by, the current series.

*P. Dierckx: *Curve and surface fittings with splines*
*J. H. Wilkinson: *The algebraic eigenvalue problem*
*I. Duff, A. Erisman, & J. Reid: *Direct methods for sparse matrices*
*M. J. Baines: *Moving finite elements*
*J. D. Pryce: *Numerical solution of Sturm-Liouville problems*
K. Burrage: *Parallel and sequential methods for ordinary differential equations*
Y. Censor & S.A. Zenios: *Parallel optimization: theory, algorithms and applications*
M. Ainsworth, J. Levesley, W. Light & M. Marletta: *Wavelets, multilevel methods and elliptic PDEs*
W. Freeden, T. Gervens, & M. Schreiner: *Constructive approximation on the sphere: theory and applications to geomathematics*
Ch. Schwab: *p- and hp- finite element methods: theory and applications to solid and fluid mechanics*
J.W. Jerome: *Modelling and computation for applications in mathematics, science, and engineering*
Alfio Quarteroni & Alberto Valli: *Domain decomposition methods for partial differential equations*
G. E. Karniadakis & S. J. Sherwin: *Spectral/hp element methods for CFD*
I. Babuška & T. Strouboulis: *The finite element method and its reliability*
B. Mohammadi & O. Pironneau: *Applied shape optimization for fluids*
S. Succi: *The Lattice Boltzmann Equation for fluid dynamics and beyond*
P. Monk: *Finite element methods for Maxwell's equations*
A. Bellen & M. Zennaro: *Numerical methods for delay differential equations*
J. Modersitzki: *Numerical methods for image registration*

Numerical Methods for Image Registration

Jan Modersitzki

Institute of Mathematics, University of Lübeck

OXFORD
UNIVERSITY PRESS

Great Clarendon Street, Oxford OX2 6DP

Oxford University Press is a department of the University of Oxford.
It furthers the University's objective of excellence in research, scholarship,
and education by publishing worldwide in

Oxford New York

Auckland Bangkok Buenos Aires Cape Town Chennai
Dar es Salaam Delhi Hong Kong Istanbul
Karachi Kolkata Kuala Lumpur Madrid Melbourne Mexico City Mumbai
Nairobi São Paulo Shanghai Taipei Tokyo Toronto

Published in the United States
by Oxford University Press Inc., New York

First published 2004

A catalogue record for this title is available from the British Library

Library of Congress Cataloging in Publication Data

(Data available)

ISBN 0 19 852841 8

10 9 8 7 6 5 4 3 2 1

Typeset by Newgen Imaging Systems (P) Ltd., Chennai, India
Printed in Great Britain
on acid-free paper by
Biddles Ltd. www.biddles.co.uk

ACKNOWLEDGEMENTS

Thanks to Bernd Fischer for making me wiser.

Thanks to Oliver Schmitt for making me see.

Thanks to Stephen Keeling for making me speak.

Thanks to Claudia Kremling for making me happy.

Also thanks to Martin Böhme, Matthias Bolten, Lutz Dümbgen, Jan Ehrhardt, Andre Folkers, Jan-Michael Frahm, Stefan Heldmann, Oliver Mahnke, Markus von Oehsen, Nils Papenberg, Hauke Prenzel, Swen Rist, Dirk Rutsatz, Bodo Siebert, Stefan Wirtz, and Maike Wolf.

CONTENTS

1

INTRODUCTION

1.1 The problem

In image processing one is often interested not only in analyzing one image but in comparing or combining the information given by different images. For this reason, *image registration* is one of the fundamental tasks within image processing. The task of image registration is to find an *optimal geometric transformation* between *corresponding* image data. In practice, the concrete type of the *geometric transformation* as well as the notions of *optimal* and *corresponding* depend on the specific application.

Image registration is a problem often encountered in many application areas like, for example, astro- and geophysics, computer vision, and medicine. For an overview, see, e.g., Brown (1992), Maintz & Viergever (1998), Maurer & Fitzpatrick (1993), van den Elsen et al (1993), and references therein.

Here, we focus on medical applications. In the last two decades, computerized image registration has played an increasingly important role particularly in medical imaging. Registered images are now used routinely in a multitude of different applications such as the treatment verification of pre- and post-intervention images and time evolution of an injected agent subject to patient motion. Image registration is also useful to take full advantage of the complementary information coming from multimodal imagery, like, for example, computer tomography (CT) and magnetic resonance imaging (MRI).

However, the interpretation of medical images and of the registration result typically requires expert knowledge. For this reason we use the simple test images shown in Fig. 1.1 throughout this book, where even a non-expert has an intuitive understanding of the outcome of a registration procedure. Here, two (modified) X-ray images of a human hand are depicted; see also Amit (1994). Note that a comparison of these two images would be much easier if they could have been aligned.

Another typical example is depicted in Fig. 1.2. In this application, magnetic resonance images of a female breast are taken at different times (images from Bruce Daniel, Lucas Center for Magnetic Resonance Spectroscopy and Imaging, Stanford University). The first image shows an MRI section taken during the so-called wash-in phase of a radiopaque marker and the second image shows the analogous section during the so-called wash-out phase. A comparison of these two images indicates a suspicious region in the upper part of the images. This region can be detected easily since the images have been registered: tissue located at a certain position in the wash-in image is related to tissue in the same position

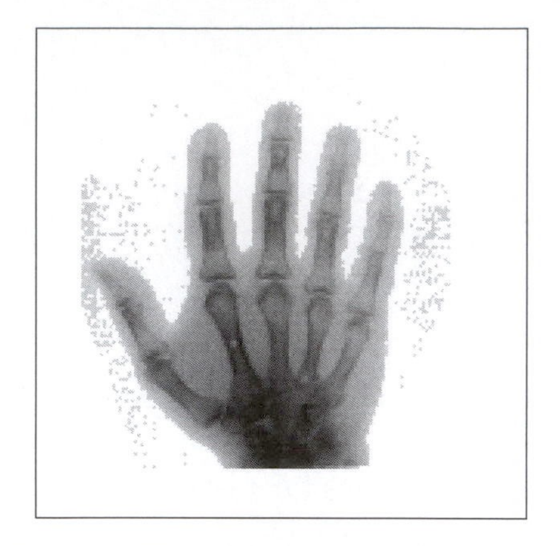

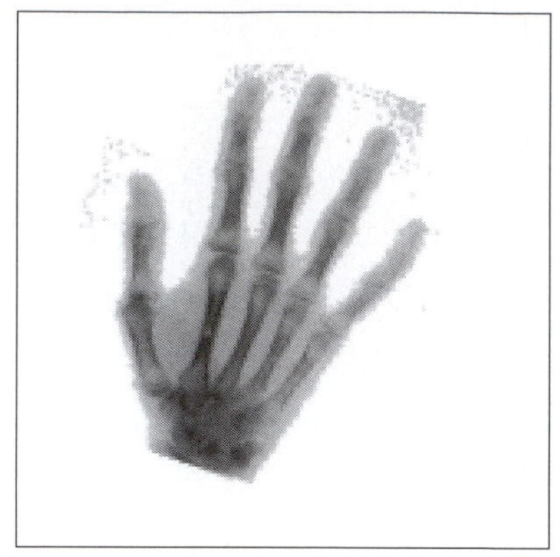

FIG. 1.1 Two different images of human hands; images from Amit (1994).

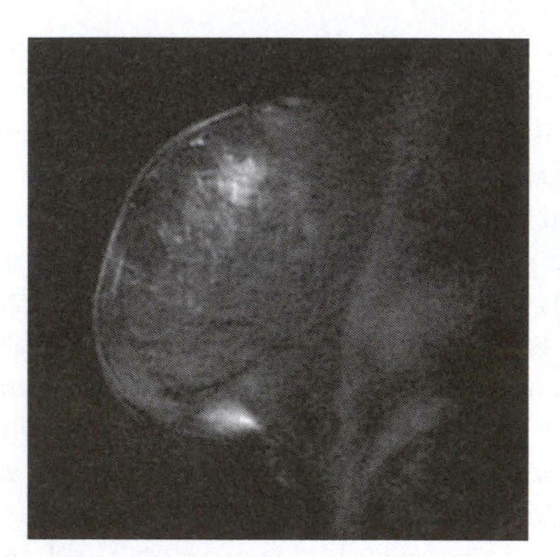

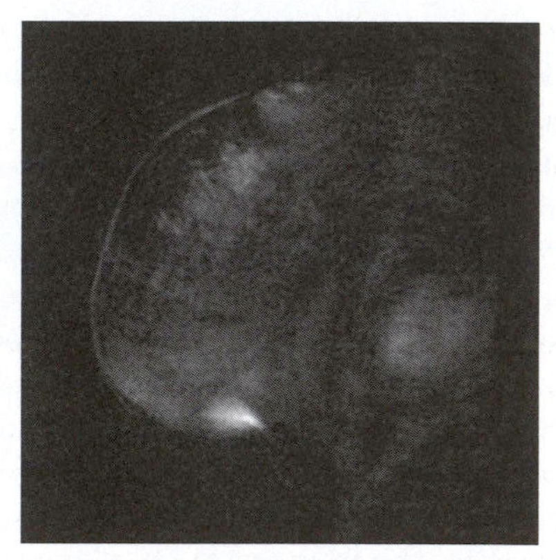

FIG. 1.2 Magnetic resonance images of a female breast, LEFT: during the wash-in phase, RIGHT: during the wash-out phase.

in the wash-out phase. Without registration, it would be much harder to decide whether an increase of brightness is related to contrast uptake or to movement of the patient, e.g., due to breathing and/or heart beat. In this application, *rigid registration* (i.e., a registration based on rotations and translations) does not provide a sufficient solution. A non-linear transformation is necessary to correct the local differences in the images.

Histological sectioning processes represent a further important source of registration problems. Here, two-dimensional slices of a three-dimensional object are produced. These two-dimensional views may serve as a basis for further image processing on a very fine scale using a microscope. But often one is interested not only in the microscopic data related to the two-dimensional views but in

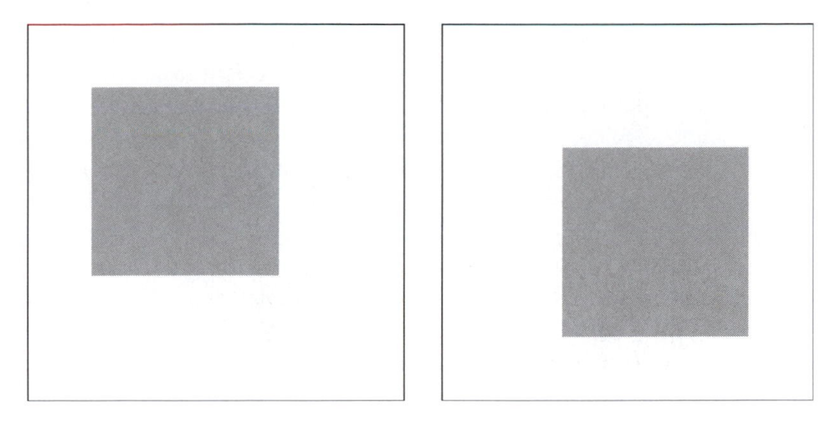

FIG. 1.3 Two squares, LEFT: reference image, RIGHT: template image.

the relation to the three-dimensional object. Typically, the sectioning process introduces individual distortions of the two-dimensional views. Thus, a direct reconstruction of the microscopic data is not possible, in general, and there is a demand for high-resolution registration procedures. A detailed description of this registration problem is given in Chapter 2.

The image registration problem can be phrased in only a few words: given a *reference* and a *template* image, find a suitable *transformation* such that the transformed template becomes *similar* to the reference. However, though the problem is easy to state, it is hard to solve. The main reason is that the problem is *ill-posed*. Small changes of the input images can lead to completely different registration results. Moreover, the solution may not be unique. Figure 1.3 illustrates a simplified situation. Suppose we have to register the reference and template images depicted in Fig. 1.3, where, for simplicity reasons, we only allow for rigid transformations, i.e., rotations and transformations. We immediately find several solutions: a pure translation, a rotation of 180 degrees, a rotation of 90 degrees followed by a translation, etc. These solutions are equivalent and without additional knowledge it is not possible to decide which one to use.

Another reason for making image registration such a challenging task is the fact that each application has its own demands with respect to the meaning of *similar* and *suitable*. For example, for the registration of X-ray images of bones, one may consider only rigid transformations, whereas for the registration of breast MRI scans one has to consider non-rigid transformations.

In the example of the hand (cf., Fig. 1.1), one could imagine using a deformable image model, based on the deformation properties of muscles, fat, bones, joints, and so forth. However, this model requires additional knowledge, e.g., the location of bones etc., which is in general not available in image registration. To make the problem even worse, image registration is often intended to supply this type of knowledge. A typical application of image registration is to map a database connected to a reference onto a template image. For example, a very accurate segmentation of a reference image could be used as a basis for a template segmentation after registration.

Finally, some applications allow for time consuming computations while others demand real-time implementation. For example, shape changes of the brain (so-called *brain-shift*) during neurosurgery caused by the intervention need to be corrected numerically during the surgery, while the correction of a histological serial sectioning may run overnight.

Thus, one also has to compromise between complex modeling and accuracy on the one side and computing time and storage requirements on the other side. Therefore, it is not surprising that a huge variety of registration techniques has been proposed in the literature.

1.2 The intention

The intention of this book is to contribute to image registration from a mathematical point of view. A general approach to image registration is presented. This approach is based on a similarity measure and a regularizer. The similarity measure can be viewed as the driving force of the registration whereas the regularizer controls the transformation. This general framework enables the classification of well-known methods as well as the design of novel schemes with particular features. Moreover, it also allows for the design of very fast implementations.

The book is divided into two parts: parametric and non-parametric image registration. Though this partitioning is slightly misleading (because there are some parameters also in non-parametric registration), it provides a good starting point.

The first part summarizes and relates state of the art registration techniques with respect to the new, general approach. The schemes presented are restricted to an expansion of the desired transformation in terms of a finite number of basis functions. Thus, registration results in the computation of optimal parameters for the underlying expansion.

The second and main part demonstrates how non-parametric registration techniques can be designed on the basis of the new approach. It is shown how well-established registration schemes like *elastic* and *fluid registration* can be related. Moreover, it is also shown how to derive new schemes with particular features using this general approach. In addition, a general numerical treatment is presented.

Finally, and most importantly, we present fast, efficient, and stable implementations for the proposed registration techniques. In particular, for *diffusion registration*, an $\mathcal{O}(N)$ implementation is presented, where N denotes the number of pixels or voxels. Using this implementation, diffusion registration is at present to the best of our knowledge the fastest non-parametric registration technique.

1.3 An overview

In the first part of the book, we address the so-called *landmark*-based (cf., Chapter 4), *principal axes*-based (cf., Chapter 5), and *optimal linear* registration (cf., Chapter 6). In landmark-based registration, one locates a number of positions in the reference image and corresponding points in the template image. The transformation is obtained by minimizing the distance (typically the Euclidean or a Mahalanobis distance) between the landmarks in the reference and deformed template image. From a mathematical point of view it is irrelevant whether or not these points are anatomically meaningful (*anatomical landmarks*) and whether these points have been selected before (*fiducial markers*) or after the imaging. The main point is that image features are deduced and the computed transformation relates these features. The registration is based purely on these features further image knowledge is not taken into account. The transformation can be regularized in different ways: it may be taken from a parameterizable space (cf., Section 4.2), e.g., a polynomial or a spline space, or be characterized by an additional smoothness property (cf., Section 4.3). Feature-based registration on a linear space is a delicate matter, because of a possible singularity in the underlying Vandermonde system. Thin plate spline registration provides a remedy. Here, the transformation is not restricted to a linear space but regularized explicitly.

The technique presented in Chapter 5 is based on a similar idea. The difference now is that geometrical features, i.e., center of gravity and the principal axes of a displayed object, are used for registration. The main advantage of this approach is that these features can be deduced automatically. Thus, no user interaction is required.

In Chapter 6, image intensity-based distance measures are introduced. In particular, we consider the *sum of squared differences* (cf., Section 6.1), a *correlation*-based distance measure (cf., Section 6.3), and *mutual information* (cf., Section 6.4). The transformation is restricted to a parameterized space and the registration can be obtained by minimizing the image distance. For presentation reasons, we restrict the transformations to an affine linear space. However, this technique carries over to all sets of parameterizable transformations. Note that the minimization of a distance measure over a parameterized spline space is probably the most commonly used registration technique in today's image registration.

The second part of the book is devoted to non-parametric image registration. The transformation is no longer restricted to a parameterizable set. Instead, a regularizing term is used to circumvent ill-posedness and to privilege more likely solutions. It is desirable to provide different regularizers since different applications require a different treatment in general. Chapter 8 is intended as an introduction to a general approach which comprises the particular methods presented in the following chapters: namely, *elastic* (Chapter 9), *fluid* (Chapter 10), *diffusion* (Chapter 11), and *curvature* registration (Chapter 12).

The registration problem is phrased in a functional setting. The main difference with respect to the parametric case where one is looking for optimal parameters of an expansion of the transformation is that we are now simply seeking a smooth transformation. No parameters are involved in representing the transformation.

This formulation presents a unified approach to image registration based on the regularized minimization of a distance measure. As has already been pointed out, image registration is an ill-posed problem. Thus, regularization is essential and inevitable. The regularizer is needed and used to pick out the most likely solution. Moreover, regularization can be used to supply additional knowledge. Actually, it is the regularizer which distinguishes the methods presented.

In Chapter 9 we comment on elastic registration in detail. The deformation of an elastic body is explained using linear elasticity theory. Note that a linear elasticity model restricts the transformation to small spatial variations. Hence, for some applications, e.g., the registration of images obtained from different human brains, elastic deformations are too restrictive by far. Thus, we also comment on fluid registration (cf., Chapter 10), where the intention is to mimic a flow of fluid in a certain sense. This approach allows for large deformations.

Diffusion registration (cf., Chapter 11) is another registration technique which can easily be derived from the general frame presented in Chapter 8. Here, we borrow the well-known optical flow regularizer for image registration. Moreover, we relate this technique to another popular non-parametric registration technique: Thirion's so-called *demons registration*. The main advantages of this regularizer is that it allows for a fast implementation. Thus, this scheme is attractive for large-scale two-dimensional and three-dimensional registration problems.

Finally, we illustrate how to use this general framework to design regularizers with particular features. In Chapter 12 we discuss a novel non-parametric registration technique which relies on a curvature-based penalizing term. This approach not only provides smooth solutions but also allows for automatic rigid alignment. Thus, in contrast to other popular registration schemes, the pre-registration step becomes redundant.

A further important aspect of image registration is the acceptance of a method in a practical environment. For this reason, robust and efficient schemes together with high-quality results become key issues. For parametric registration standard minimization techniques, like, for example, Gauss–Newton or Levenberg–Marquardt techniques, are used commonly and, with some additional effort, can even be applied to the minimization of mutual information.

Unfortunately, the computation of a numerical solution for a non-parametric registration is not straightforward. Thus, we also present algorithms for all four non-parametric registration schemes. The corresponding Euler–Lagrange equations serve as a starting point for all implementations. After an appropriate discretization, this approach leads to high-dimensional linear systems of equations which have to be solved numerically. Though this dimension can get quite large, e.g., $N = 52\,428\,800$ for the registration of the histological serial sectioning presented in Section 9.10, the underlying systems are highly structured. The structure enables the design of efficient solution schemes. The complexity of the devised schemes is $\mathcal{O}(N \log N)$ for elastic, fluid, and curvature registration and $\mathcal{O}(N)$ for diffusion registration, where N denotes the number of image pixels or voxels or whatever. These fast solution schemes form the backbone of our implementations.

Though the proposed techniques are very general and can be used for various applications, we focus on one particular application: the reconstruction of a histological serial sectioning of a human brain (cf., Chapter 2). However, additional examples are presented to illustrate the advantages and disadvantages of the various techniques. The hand images (cf., Fig. 1.1) serve as the example in this respect.

2

THE HUMAN NEUROSCANNING PROJECT

The **H**uman **N**euro**S**canning **P**roject (HNSP) was initiated by O. Schmitt; cf., Schmitt (2001). The goal of this project is a three-dimensional reconstruction down to particular neurons of a whole human brain based on microscopic modalities. These data should then be used as the basic structure for the integration of functional data based on stochastic mapping and later on for modeling and simulation studies of a virtual brain; see Schmitt (2001) for details.

In order to locate the spatial position of the neurons, the post mortem brain from a 55 year old male human voluntary donor was prepared in several steps. The brain was fixed in a neutral buffered formaldehyde solution for three months. An MRI scan of the brain was produced after fixation. Dehydration and embedding of the brain in paraffin required three further months.

This preparatory work was followed by sectioning the brain in 20 μm thick slices (about 5000 for this brain) using a sliding microtome; cf., Fig. 2.1. Before each slicing step a high-resolution episcopic image (1352 × 1795 pixels, three colors) was taken; cf., Fig. 2.2.

Figure 2.3 displays a tissue slice after sectioning. The tissue slice was then stretched in warm water at 55°C. Thereafter, it was transferred onto a microscopic slide and dried overnight. After drying, the sections were stained in gallocyanin chrome alum and mounted under cover-glasses.

A specialized light microscope (LMAS) with an extraordinarily large object range of 250×250 mm^2 is used to visualize all cells of the large tissue sections; cf., Fig. 2.2. Different neuronal entities were analyzed on different structural scales, i.e., from macroscopic details down to the cellular level; see Schmitt (2001) for the image processing.

In order to relate the microscopic data to a macroscopic view of the slide and to recover the geometrical deformation of the tissue introduced by the various sectioning steps, transparent flat bed scans (FBS) of the slides were produced; cf., Fig. 2.3. Using a resolution of 2032 parts per inch (ppi) in an 8 Bit gray-scale mode the digitized images range between 5000 × 2000 and 11000 × 7000 pixels (about 196 MBytes storage for the largest scan).

The sectioning, stretching, and drying processes are essential to produce high-quality slides for further microscopic analysis; cf., Schmitt (2001). However, these methodological steps produce non-linear (i.e., not necessarily linear) deformations of the tissue. A deformed tissue section is visible in Fig. 2.3. These deformations prohibit a direct three-dimensional reconstruction of the data. Image registration techniques are required to by-pass these problems.

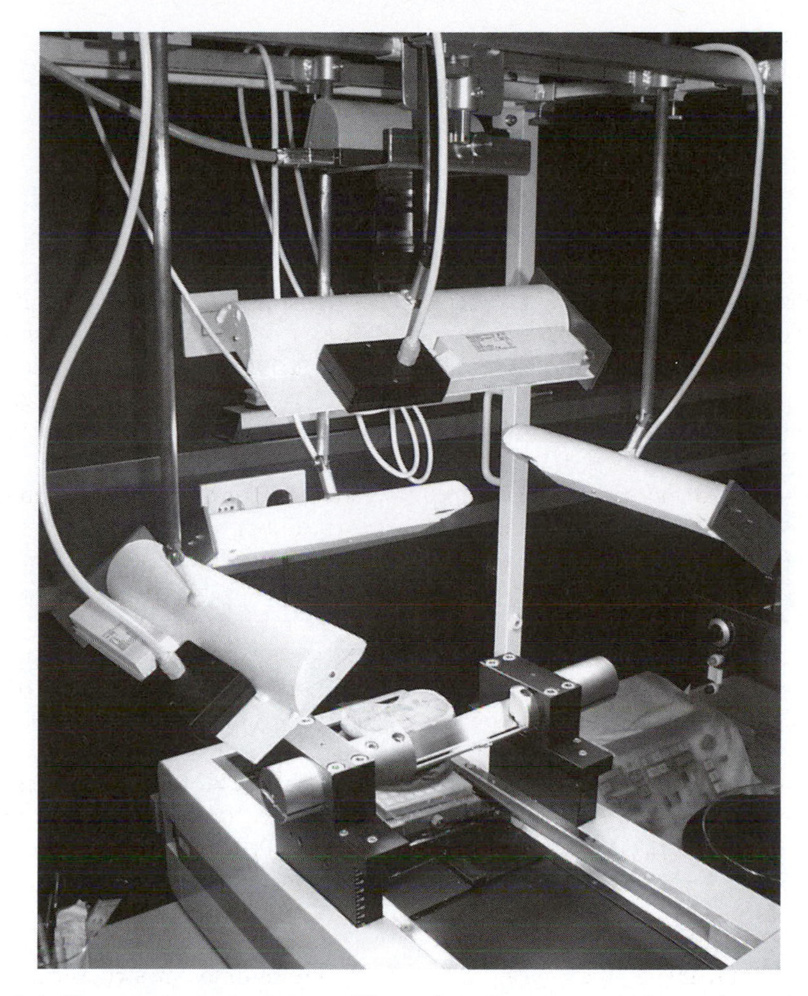

FIG. 2.1 The sectioning process. The episcopic camera is located at the top, the sliding microtome and the paraffin embedded brain are located at the bottom.

Figure 2.4 displays the arbitrarily chosen section 3799 (reference) and section 3800 (template) as well as the difference between the non-registered, the linearly registered, and the non-linearly registered sections. As is apparent from this figure, non-linear registration techniques may provide effective tools for the reconstruction of geometrical deformations.

In this phase of the HNSP the images were re-sampled to 1024×1024 pixels in order to limit the memory requirements and the computation times while providing adequate resolution. A straightforward re-sampling based on bi-linear interpolation was used.

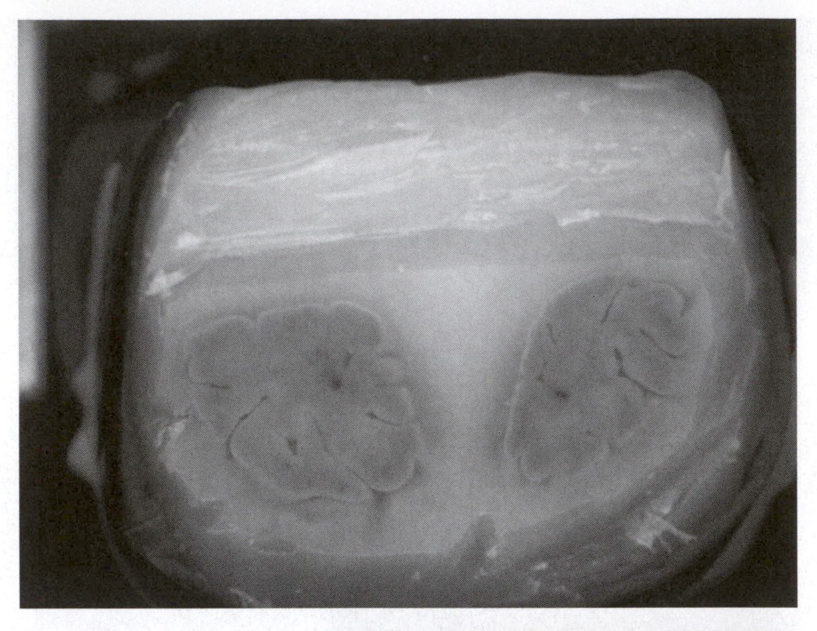

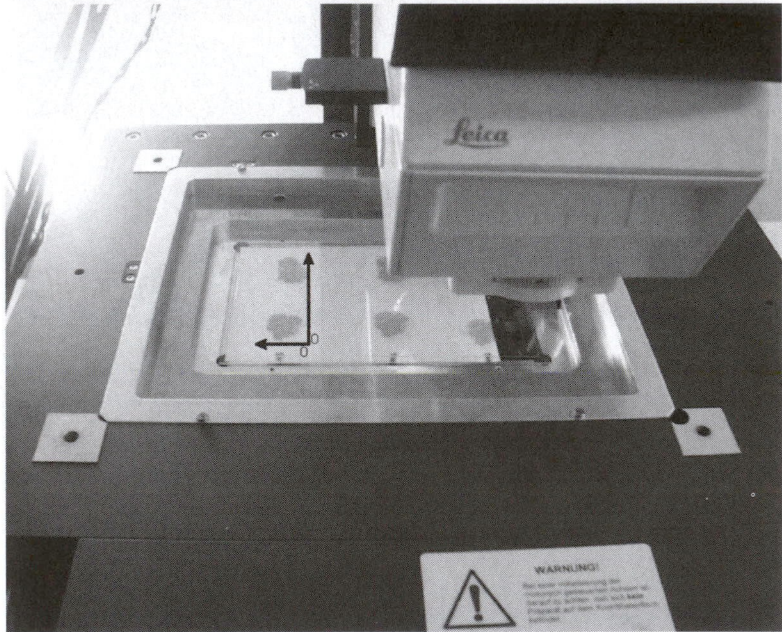

FIG. 2.2 TOP: Episcopic image of the paraffin embedded brain before sectioning section 100; BOTTOM: slide under a particular light microscope.

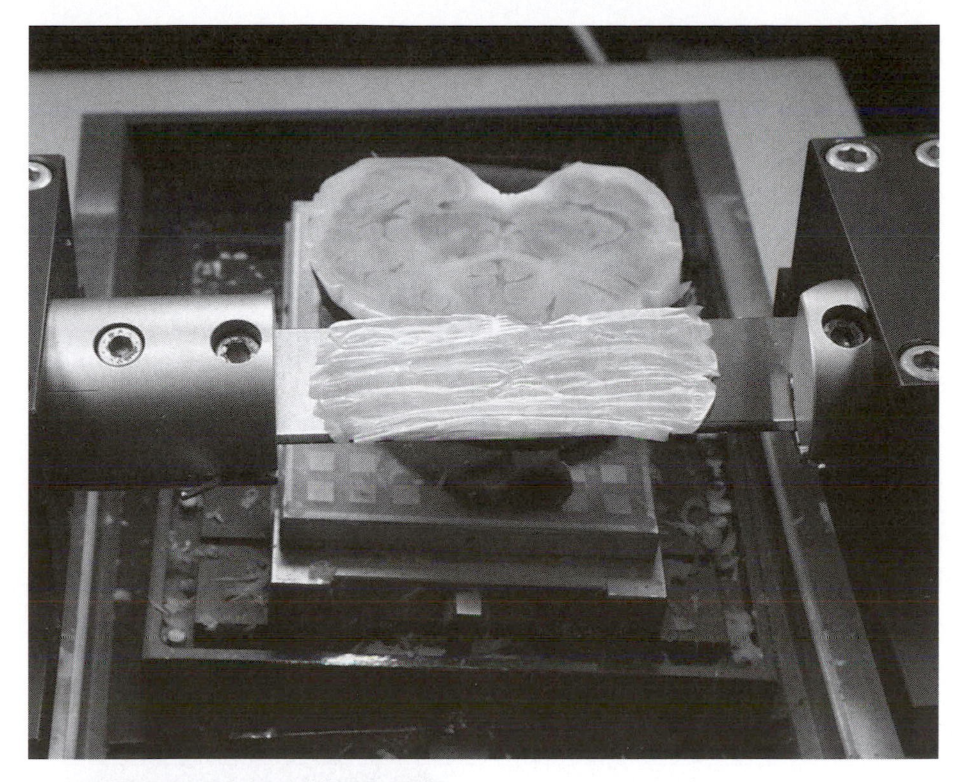

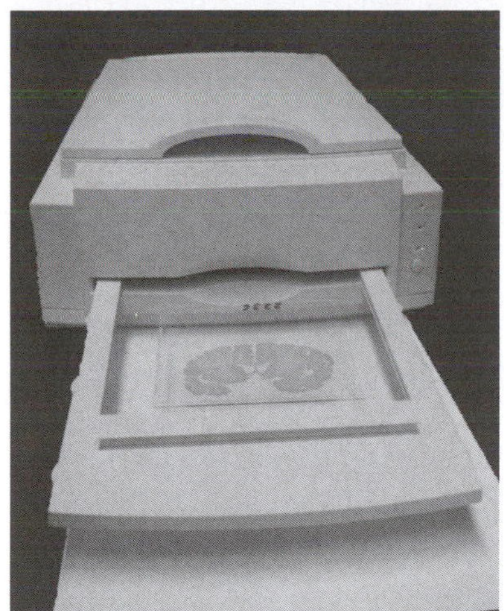

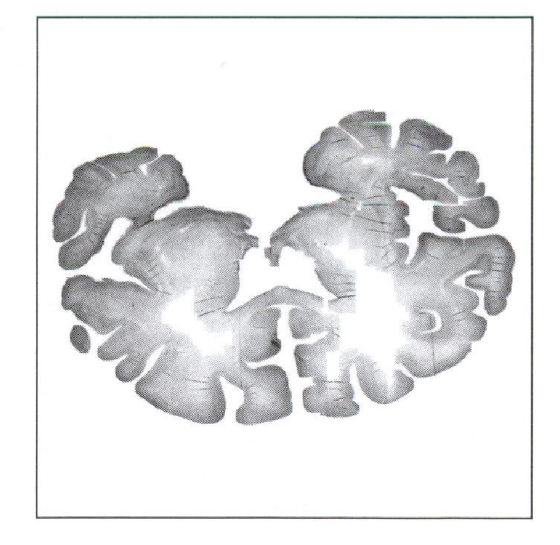

FIG. 2.3 TOP: microtome with tissue section on top of the blade, BOTTOM LEFT: flat bed scanner, BOTTOM RIGHT: transparent flat bed scan of the tissue section after fixation.

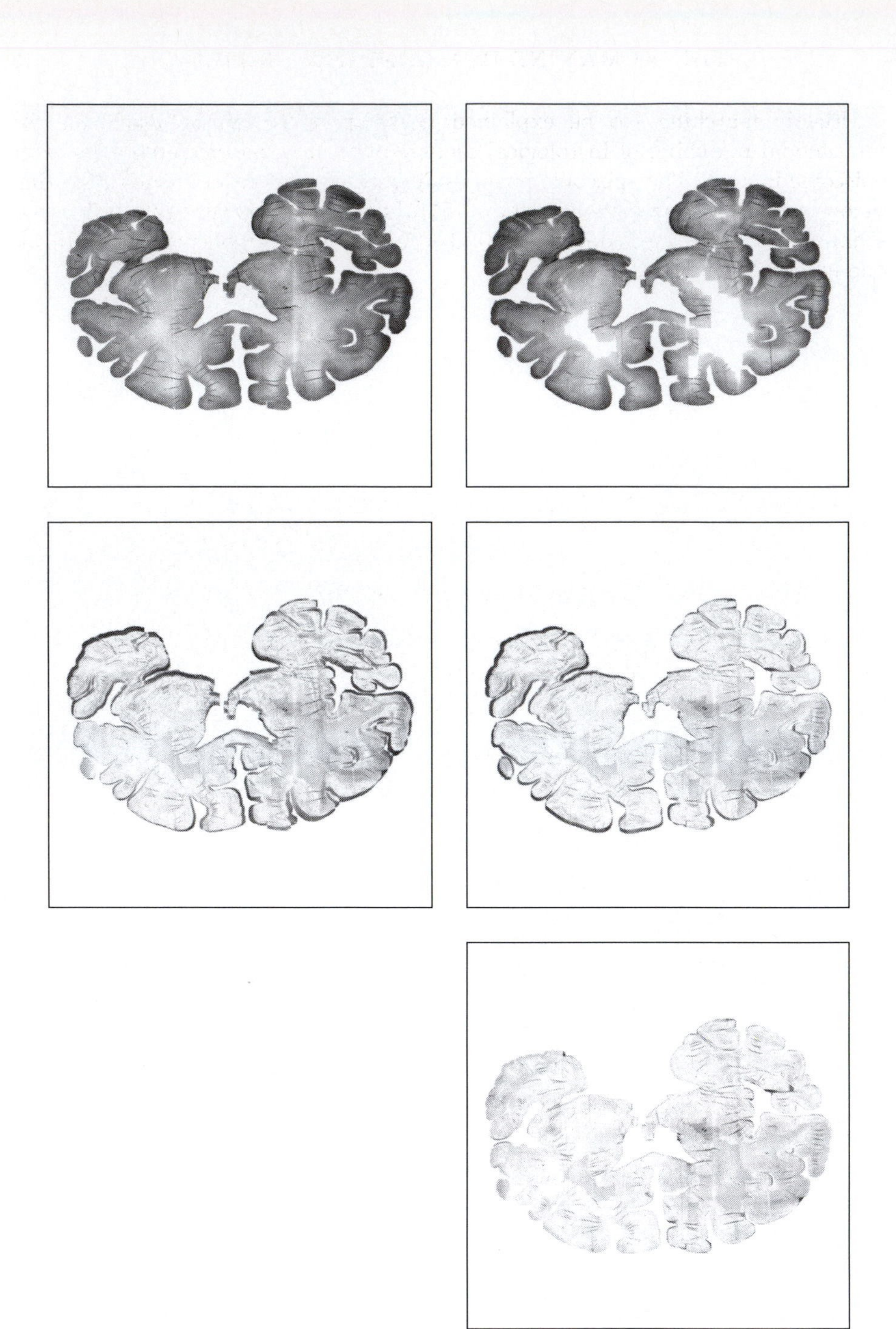

FIG. 2.4 Sections 3799 (TOP LEFT) and 3800 (TOP RIGHT) as well as the difference images non- (MIDDLE LEFT), linearly (MIDDLE RIGHT), and non-linearly registered.

Elastic matching (to be explained in Chapter 9) can be used also for multimodal matching of histological sections with non-deformed MRI scans or episcopic images. The episcopic images might be derived from image processing before sectioning the embedded brain. The uncompressed amount of flat bed scanned data was approximately 700 GBytes + 40 GBytes episcopic data for one human brain.

3

THE MATHEMATICAL SETTING

3.1 General remarks

3.1.1 *Introduction to images*

Here, we use a very simple model of an image, a mapping which assigns every spatial point x belonging to a certain set $\Omega \subset \mathbb{R}^d$ a *gray value* $b(x)$. Consequently, an image is viewed as a mapping. The dimension of its spatial domain is denoted by $d \in \mathbb{N}$ throughout this book.

Definition 3.1 *Let* $d \in \mathbb{N}$*. A function* $b : \mathbb{R}^d \to \mathbb{R}$ *is called a* d-dimensional image, *if*

1. b *is compactly supported,*
2. $0 \leq b(x) < \infty$ *for all* $x \in \mathbb{R}^d$*,*
3. $\int_{\mathbb{R}^d} b(x)^k dx$ *is finite, for* $k > 0$*.*

The set of all images is denoted by

$$\mathrm{Img}(d) := \left\{ b : \mathbb{R}^d \to \mathbb{R} \mid b \textit{ is } d\textit{-dimensional image} \right\}.$$

Here, the term image refers to a d-dimensional light-intensity function b; the value $b(x)$ gives the intensity of the image at a spatial position $x \in \mathbb{R}^d$. Although Definition 3.1 (1) looks restrictive at first glance, it appears to be natural for the registration of the histological sections within the HNSP. A human brain is bounded and the brain matter is completely embedded in paraffin wax. However, it might become crucial in other applications. Definition 3.1 (3) conforms naturally to the boundedness of light energy.

For all the applications we are interested in, the images are given in terms of discrete data and the intensity function has to be interpolated by some interpolation scheme; cf., Section 3.1.3. Thus, by choosing an appropriate interpolation scheme, we may even assume that the images are arbitrarily smooth. This assumption enables us to exploit fast numerical schemes which typically depend on second order derivatives.

One may also be interested in q-colored images, i.e., mappings $c : \mathbb{R}^d \to \mathbb{R}^q$, $c = (c^1, \dots, c^q)$, where $c^\ell \in \mathrm{Img}(d)$ for $\ell = 1, \dots, q$. The following concepts apply to any component of the q-colored image and can thus be extended in a straightforward way.

3.1.2 *From continuous to discrete images*

We now introduce *digital images* as a model for the discrete output of an imaging device. For an image $b \in \mathrm{Img}(d)$ we assume that the support of the image b is contained in a domain Ω, where for ease of presentation we assume

$$\Omega :=]0,1[^d \subset \mathbb{R}^d,$$

and denote the boundary of Ω by $\partial\Omega$.

Definition 3.2 *Let $d \in \mathbb{N}$, $\Omega :=]0,1[^d$, and $n_1, \dots, n_d \in \mathbb{N}$ be some given numbers. The points*

$$x_{j_1,\dots,j_d} = (x_{j_1}, \dots, x_{j_d})^\top \in \Omega \cup \partial\Omega, \tag{3.1}$$

where $1 \le j_\ell \le n_\ell$ and $1 \le \ell \le d$, are called grid points. *The array*

$$X := (x_{j_1,\dots,j_d})_{\substack{1 \le j_\ell \le n_\ell \\ \ell=1,\dots,d}} \in \mathbb{R}^{n_1 \times \cdots \times n_d} \tag{3.2}$$

is called the grid matrix.

Let $N := n_1 \cdots n_d$ and let the numbers $j \in \mathbb{N}$ $(1 \le j \le N)$ and $(j_1, \dots, j_d) \in \mathbb{N}^d$ $(1 \le j_\ell \le n_\ell,\ 1 \le \ell \le d)$ be related by the one-to-one lexicographical ordering, $j = \sum_{\nu=1}^{d-1}(j_{\nu+1} - 1)\prod_{\mu=1}^{\nu} n_\mu + j_1$. The vector $\vec{X} := (x_j)_{j=1,\dots,N} \in \mathbb{R}^N$, where $x_j = x_{j_1,\dots,j_d}$, is called the grid vector. *The set $\Omega_d := \{ x_j, j = 1, \dots, N \}$ is called an $n_1 \times \cdots \times n_d$* grid.

In connection with the discretization of the partial differential equations to be discussed in Part II, we now introduce three different grids which are of particular interest.

Definition 3.3 *Let $d \in \mathbb{N}$, $\Omega :=]0,1[^d$, and $n_1, \dots, n_d \in \mathbb{N}$ be some given numbers. For $1 \le j_\ell \le n_\ell$, $1 \le \ell \le d$, let*

$$x_j^D := \left(\frac{j_1}{n_1 + 1}, \dots, \frac{j_d}{n_d + 1} \right)^\top, \tag{3.3}$$

$$x_j^N := \left(\frac{2j_1 - 1}{2n_1}, \dots, \frac{2j_d - 1}{2n_d} \right)^\top, \tag{3.4}$$

$$\text{and} \quad x_j^p := \left(\frac{j_1 - 1}{n_1}, \dots, \frac{j_d - 1}{n_d} \right)^\top \tag{3.5}$$

be grid points. The corresponding grids Ω_D, Ω_N, and Ω_p are called Dirichlet, Neumann, *and* periodic grids, *respectively.*

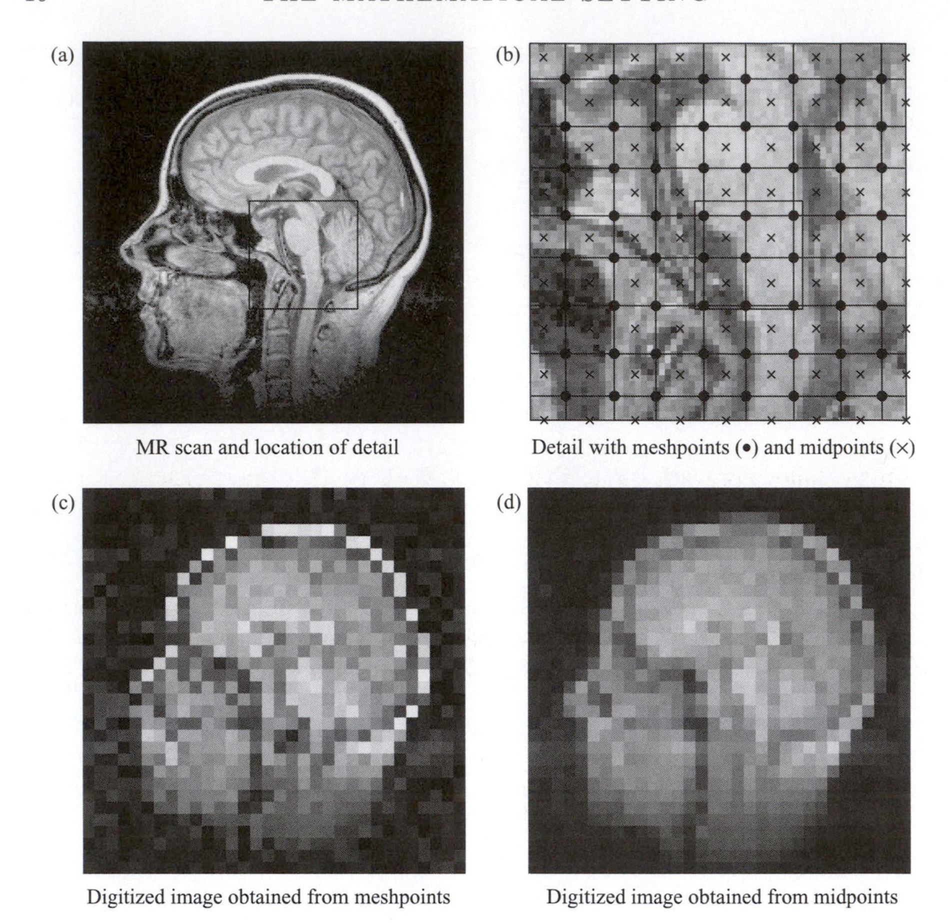

FIG. 3.1 (a) Image, (b) mesh- and midpoints in a detail of the image, (c) mesh and (d) midpoint digitized images (32 × 32 pixels).

For the discretization of a continuous image b on a discrete grid Ω_d two models are common; see also Fig. 3.1 for an illustration.

Meshpoint model:

$$b_j^{\bullet} := b_{j_1,\ldots,j_d}^{\bullet} := b(x_j) \quad \text{for all} \quad x_j \in \Omega_d. \tag{3.6}$$

Midpoint model:

$$b_j^{\times} := b_{j_1,\ldots,j_d}^{\times} := \int_{c_j} b(x)dx \quad \text{for all} \quad j = 1, \ldots, N. \tag{3.7}$$

Here, c_j denotes a small region with center x_j.

For both models we define $b(x) = 0$ for $x \notin \Omega$.

Definition 3.4 *Let $d \in \mathbb{N}$, $b \in \mathrm{Img}(d)$, and let Ω_d be a $n_1 \times \cdots \times n_d$ grid; cf., Definition 3.2. The arrays*

$$B^\bullet := (b^\bullet_{j_1,\ldots,j_d})_{\substack{j_\ell=1,\ldots,n_\ell,\\ \ell=1,\ldots,d}}, \quad B^\times := (b^\times_{j_1,\ldots,j_d})_{\substack{j_\ell=1,\ldots,n_\ell\\ \ell=1,\ldots,d}} \in \mathbb{R}^{n_1\times\cdots\times n_d},$$

where $b^\bullet_j$ and $b^\times_j$ are obtained respectively from the mesh- or the midpoint approach, are called d-dimensional digital images.

If the distinction between mesh- and midpoint discretization is not of importance, we may simply denote a d-dimensional discrete image by $B \in \mathbb{R}^{n_1\times\cdots\times n_d}$.

In the following, we assume that a digital image B is an appropriate model for the output of an imaging device. This is, of course, a considerable simplification. Different devices yield different outputs. Moreover, the output depends on additional parameters, e.g., light.

Many imaging devices, e.g., CCD cameras or flat bed scanners, perform a further image processing step by also digitizing the image values. In particular for the scans of the two-dimensional HNSP images we end up with values $\widetilde{b}_i \in \{0,\ldots,255\}$, where

$$\widetilde{b}_j = \begin{cases} 0, & b_j < 0, \\ 255, & b_j > 255, \\ [b_j], & \text{else.} \end{cases}$$

Here and later,

$$\lfloor x \rfloor := \max\{j \in \mathbb{Z} \mid j \le x\}, \tag{3.8}$$

$$\lceil x \rceil := \min\{j \in \mathbb{Z} \mid j \ge x\}, \tag{3.9}$$

$$[x] := \min\{j \in \mathbb{Z} \mid j - 0.5 \le x < j + 0.5\}. \tag{3.10}$$

We also use this notation for vectors, e.g., $[x] = ([x_1],\ldots,[x_d])^\top$ for $x \in \mathbb{R}^d$.

3.1.3 *From discrete to continuous images*

Given a discrete image B one may also want to assign image values at spatial positions that are not necessarily grid points. In other words, we are looking for an *interpolation scheme* $\mathcal{I}$,

$$\mathcal{I} : \mathbb{R}^{n_1 \times \cdots \times n_d} \times \mathbb{R}^d \to \mathbb{R}.$$

For an image B, the scheme assigns an intensity value $\mathcal{I}(B, x)$ to any position $x \in \mathbb{R}^d$.

We discuss different variants; see also Fig. 3.2 for a comparison. For ease of presentation, we focus on an $n_1 \times \cdots \times n_d$ periodic grid. The extension to other grids is straightforward. For $x \notin \Omega$ we set $\mathcal{I}(B, x) := 0$. Let $n = (n_1, \ldots, n_d)$ and $x \in \Omega$ be an arbitrary point and $[x]$ the corresponding closest grid point and $\lfloor x \rfloor$ the corresponding closest grid point smaller then x, i.e.,

$$[x] = \left(\frac{[x_1 n_1]}{n_1}, \ldots, \frac{[x_d n_d]}{n_d} \right)^\top \quad \text{and} \quad \lfloor x \rfloor = \left(\frac{\lfloor x_1 n_1 \rfloor}{n_1}, \ldots, \frac{\lfloor x_d n_d \rfloor}{n_d} \right)^\top.$$

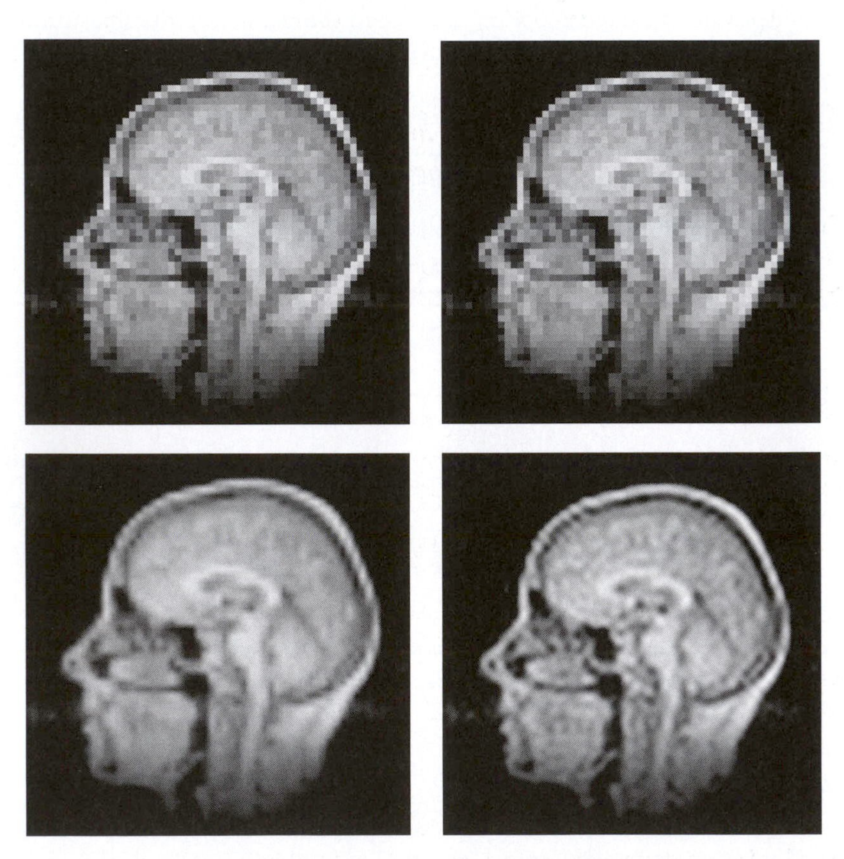

Fig. 3.2 Top left: Digital image 59×59 pixels, top right: next-neighbor interpolation, bottom left: bi-linear interpolation, and bottom right: interpolation based on cosine basis functions.

Local methods

1. *Next-neighbor interpolation*

$$\mathcal{I}_{\mathrm{NN}}(B,x) := B([x]). \tag{3.11}$$

   Note that the function $\mathcal{I}_{\mathrm{NN}}(B,\,\cdot\,)$ is in general not continuous.

2. *d-linear interpolation*

$$\begin{aligned}&\mathcal{I}_{\mathrm{linear}}(B,x)\\ &:= \sum_{k\in\{0,1\}^d} B\left(\frac{\lfloor n_1x_1\rfloor + k_1}{n_1},\ldots,\frac{\lfloor n_dx_d\rfloor + k_d}{n_d}\right)\\ &\times \prod_{j=1}^{d}\left(\tfrac{(-1)^{k_j}}{n_j}(\lfloor x_jn_j\rfloor + 1 - k_j - x_jn_j)\right).\end{aligned} \tag{3.12}$$

   Note that the function $\mathcal{I}_{\mathrm{linear}}(B,\,\cdot\,)$ is continuous but in general not continuously differentiable.

   As a compromise between computational effort and accuracy, we typically exploit bi- ($d = 2$) and tri-linear ($d = 3$) interpolation schemes here.

Global methods

Let N be the number of grid points and $(\psi_j, j = 1,\ldots,N)$ be a set of basis functions such that the interpolation problem

$$\sum_{j=1}^{N} \alpha_j\psi_j(i) = B(i), \quad i \in \Omega_d$$

has a unique solution for any $B \in \mathbb{R}^{n_1\times\cdots\times n_d}$. For a given discrete image B we define

$$\mathcal{I}_{\psi_1,\ldots,\psi_N}(B,x) := \sum_{j=1}^{N} \alpha_j\psi_j(x). \tag{3.13}$$

Typical choices of basis function are tensor products of Lagrange interpolation polynomials, sinc functions (see, e.g., Shannon (1949)), B-splines (see, e.g., Thévenaz et al (2000)), a collection of sine and cosine functions, a collection of cosine functions, or wavelets (see, e.g., Aldroubi & Gröchening (2001)).

3.2 Mathematical notation

For later reference, we introduce some general notation. Let $\Omega \subset \mathbb{R}$ be an open set. By $L_2(\Omega)$ we denote the set of squared integrable functions,

$$L_2(\Omega) := \left\{ f : \Omega \to \mathbb{R} \mid \int_\Omega |f(x)|^2\, dx < \infty \right\}$$

with the inner product

$$\langle f, g\rangle_{L_2} := \int_\Omega f(x)g(x)\,dx.$$

Furthermore, by $\mathcal{C}^k(\Omega)$ we denote the set of k-times differentiable functions and by $\mathcal{C}_c^k(\Omega)$ we denote the set of k-times differentiable functions with compact support.

Definition 3.5 *Let $d \in \mathbb{N}$. Moreover, let $\mathcal{D} := \mathcal{C}_c^\infty(\mathbb{R}^d)$ denote the space of arbitrary often differentiable functions with compact support and $\mathcal{D}^*$ its dual. For $z \in \mathbb{R}^d$ we define the* point evaluation functional *or* Dirac δ functional *by*

$$\delta_z[\varphi] = \varphi(z) \quad \textit{for all} \quad \varphi \in \mathcal{D}.$$

Let $f \in L_2(\mathbb{R}^d)$ be such that $\int_{\mathbb{R}^d} f(x)\,dx = 1$ and for $h > 0$, let $f_h(x) := h^{-d} f(x/h)$. For any $\varphi \in \mathcal{D}$, we have

$$\lim_{h\downarrow 0} \int_{\mathbb{R}^d} f_h(x)\varphi(x)\,dx = \varphi(0) = \delta_0[\varphi].$$

Thus, as usual, we make use of the notion "δ-function" for the limit, i.e.,

$$\delta_z[\varphi] = \int_{\mathbb{R}^d} \delta_z(x)\varphi(x)\,dx = \varphi(z).$$

For the various differential operators used throughout this book, we provide the overview given in Table 3.1.

Table 3.1 *Notation for derivatives.*

∂_{x_j}	partial derivative with respect to the jth component
∇	gradient, $\nabla f = (\partial_{x_1} f, \ldots, \partial_{x_d} f)^\top$ for $f : \mathbb{R}^d \to \mathbb{R}$, $\nabla\varphi = (\partial_{x_k}\varphi_j)_{j,k} \in \mathbb{R}^{d\times d}$ for $\varphi : \mathbb{R}^d \to \mathbb{R}^d$
∇_p	gradient with respect to the variable p
div	divergence, $\operatorname{div}\varphi = \langle\nabla, \varphi\rangle_{\mathbb{R}^d} = \partial_{x_1}\varphi_1 + \cdots + \partial_{x_d}\varphi_d$ for $\varphi : \mathbb{R}^d \to \mathbb{R}^d$
∇^2	Hessian matrix $\nabla^2 f = (\partial_{x_j,x_k} f)_{j,k} \in \mathbb{R}^{d\times d}$ for $f : \mathbb{R}^d \to \mathbb{R}$,
Δ	Laplace operator, $\Delta f = \partial_{x_1,x_1} f + \cdots + \partial_{x_d,x_d} f$ for $f : \mathbb{R}^d \to \mathbb{R}$, $\Delta\varphi = (\Delta\varphi_1, \ldots, \Delta\varphi_d)^\top$ for $\varphi : \mathbb{R}^d \to \mathbb{R}^d$
D^κ	$D^\kappa f = (\partial/\partial_{x_1})^{\kappa_1} \cdots (\partial/\partial_{x_d})^{\kappa_d} f$ for $f : \mathbb{R}^d \to \mathbb{R}$ and $\kappa \in \mathbb{N}^d$
$d\mathcal{F}[\varphi; \psi]$	Gâteaux derivative of a functional $\mathcal{F}$ at φ with respect to a perturbation ψ

Finally, we use the Kronecker calculus; see, e.g., Horn & Johnson (1991, §4) or Brewer (1978). In particular, for $A \in \mathbb{C}^{m\times n}$ and $B \in \mathbb{C}^{p\times q}$, we have

$$A \otimes B := \begin{pmatrix} a_{1,1}B & \cdots & a_{1,n}B \\ \vdots & \ddots & \vdots \\ a_{m,1}B & \cdots & a_{m,n}B \end{pmatrix} \in \mathbb{C}^{mp\times nq}.$$

The identity matrix is denoted by $I_n \in \mathbb{R}^{n\times n}$.

3.3 The registration problem

Image registration means finding a suitable spatial transformation such that a transformed image becomes similar to another one. The registration problem typically occurs when two images exhibit essentially the same object, but the object or the position of the imaging device is different for the two images. Thus, the images are spatially not *aligned* or not *registered*, i.e., there is no direct spatial correspondence between them. Typically this problem occurs if the images are taken from different perspectives, times, or imaging devices.

Usually, one of the images is viewed as a *reference* R and the other one as a deformable *template* T or *study* . Ideally, we are looking for a *transformation* $\varphi : \mathbb{R}^d \to \mathbb{R}^d$, such that the reference image R and the deformed template image T_φ are similar, where

$$T_\varphi(x) := T \circ \varphi(x) = T(\varphi(x)). \tag{3.14}$$

In many practical applications the problem is even harder. This is because the reference and template images show different but similar objects. Typical examples include the registration of images taken before and after surgery for a patient or the registration of consecutive tissue sections from a histological sectioning process. Here, the aim of the registration is to remove the *artificial* differences introduced for example by movement, but to retain the *real* differences due to the variations of the different objects.

For a mathematical treatment of the problem, similarity needs to be measured in an appropriate way. If the images are taken from different imaging devices, there might not be a correspondence between the intensities $R(x)$ and $T_\varphi(x)$ for an optimal φ. Thus, we may also allow for an additional function $g : \mathbb{R} \to \mathbb{R}$ and compare $R(x)$ and $g \circ T_\varphi(x) = g(T(\varphi(x)))$. Moreover, one may also consider distance measures which are not related to intensities but are based on the so-called *mutual information* of the images; cf., Section 6.4. Different similarity measures $\mathcal{D}$ are introduced in the forthcoming sections. The general registration problem reads as follows.

Problem 3.1 *Given a distance measure* $\mathcal{D} : \mathrm{Img}(d)^2 \to \mathbb{R}$ *and two images* $R, T \in \mathrm{Img}(d)$, *find a mapping* $\varphi : \mathbb{R}^d \to \mathbb{R}^d$ *and a mapping* $g : \mathbb{R} \to \mathbb{R}$ *such that* $\mathcal{D}(R, g \circ T \circ \varphi) = \min$.

As we will see in the following, Problem 3.1 is ill-posed. Thus, a direct approach is impossible. Different regularization approaches are discussed subsequently.

3.3.1 *Restricted transformations*

In many applications there exist implicit requirements with respect to the transformation. A typical example is that the transformation has to be smooth or even diffeomorphic. Approaches based on regularization are investigated in Part II. Explicit parametric requirements are even more popular in the literature and typical example are given below.

Rigid transformation, $\varphi(x) = Qx + b$, where $Q \in \mathbb{R}^{d\times d}$ is orthogonal with $\det Q = 1$ and $b \in \mathbb{R}^d$. The transformation is called rigid because only rotations and translations of the coordinates are permitted.

(Affine) linear transformation, $\varphi(x) = Ax + b$, $A \in \mathbb{R}^{d\times d}$ with $\det A > 0$, $b \in \mathbb{R}^d$. Note that in contrast to other authors we allow for individual scalings.

Polynomial transformation, $\varphi \in \Pi_q^d(\mathbb{R}^d)$; cf., Definition 3.6.

B-spline transformation,

$$\varphi = (\varphi_1, \dots, \varphi_d)^\top : \mathbb{R}^d \to \mathbb{R}^d, \quad \text{where} \quad \varphi_\ell \in \mathrm{Spline}_{y,q}(\mathbb{R}^d),$$

where $\mathrm{Spline}_{y,q}(\mathbb{R}^d)$ is a d-dimensional spline space spanned by B-splines of degree q with respect to the knots y_k, $k = 1, \dots, K$; cf., e.g., De Boor (1978).

For reference purposes we also present a formal definition of polynomials.

Definition 3.6 *Let* $d, q \in \mathbb{N}$ *and* $\kappa = (\kappa_1, \dots, \kappa_d) \in \mathbb{N}^d$. *The* polynomials of degree q *are defined by*

$$\Pi_q(\mathbb{R}^d) := \left\{ \psi : \mathbb{R}^d \to \mathbb{R} \mid \psi(x) = \textstyle\sum_{|\kappa|\le q} \alpha_\kappa x^\kappa, \ \alpha_\kappa \in \mathbb{R} \right\}$$

where $|\kappa| = \kappa_1 + \cdots + \kappa_d$ *and for* $x \in \mathbb{R}^d$ *we set* $x^\kappa := x_1^{\kappa_1} \cdots x_d^{\kappa_d}$. *The set of* d-dimensional polynomials of degree q *is defined by*

$$\Pi_q^d(\mathbb{R}^d) := \{ \varphi : \mathbb{R}^d \to \mathbb{R}^d \mid \varphi_\ell \in \Pi_q(\mathbb{R}^d), \ \ell = 1, \dots, d \}.$$

3.3.2 *The Lagrange and Euler frames*

For the image interpolation there are two reference frames, the Lagrange coordinates and the Euler coordinates. If we assume the transformation φ to be invertible, we can write $\tilde{x} := \varphi(x)$ or equivalently $x = \tilde{\varphi}(\tilde{x}) := \varphi^{-1}(\tilde{x})$.

 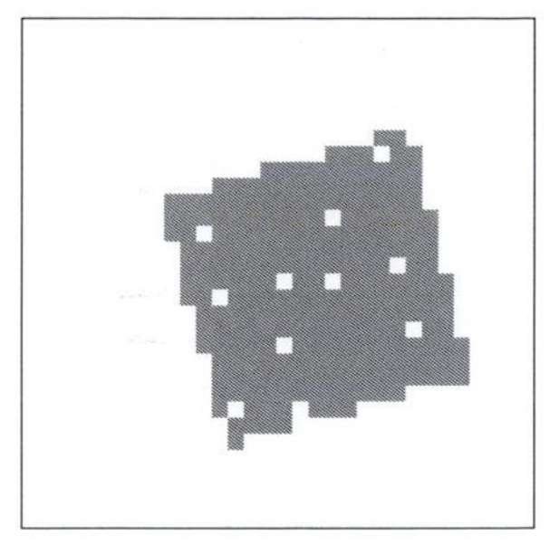 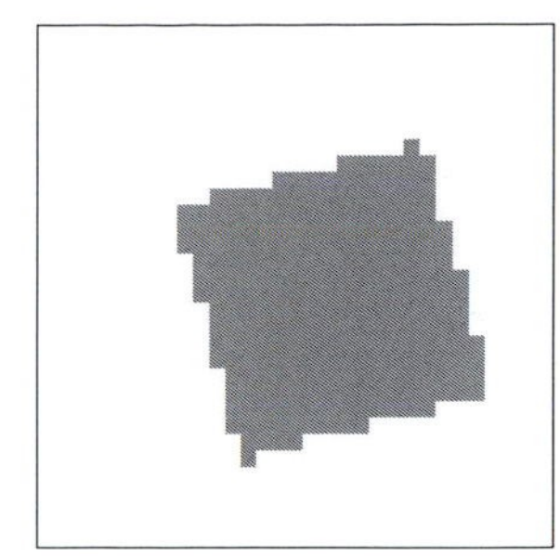

FIG. 3.3 Image and affine linear mapped image. LEFT: Input image 32×32 pixels, MIDDLE: next-neighbor interpolation using the Lagrange coordinates, RIGHT: next-neighbor interpolation using the Euler coordinates. Image has been rotated by 20 degrees.

From a computational point of view we have

$$B^{\text{Lagrange}}(\varphi(i,j)) := B(i,j) \quad \text{and} \quad B^{\text{Euler}}(i,j) := B(\varphi^{-1}(i,j)).$$

Figure 3.3 illustrates the difference between the two frames and indicates that the Euler frame is preferable. Thus, we take advantage of the Euler coordinates.

Part I

Parametric image registration

4

LANDMARK-BASED REGISTRATION

4.1 General remarks

In this first part we will discuss image registration techniques which are based on a finite set of parameters and/or a finite set of so-called *image features*. The basic idea is to determine the transformation such that for a finite number of features, any feature of the template image is mapped onto the corresponding feature of the reference image.

Typical features are, for example, "hard" or "soft" landmarks in the images; see Fig. 4.1 below for an illustration. A landmark is the location of a typically outstanding feature of an image, e.g., the tip of a finger or the point of maximal curvature. *Hard* landmarks or *prospective* landmarks are so-called *fiducial markers* which are positioned before imaging at certain spatial positions on a patient. Typically, the spatial position of these landmarks can be deduced from the images with high accuracy; see, e.g., Maurer & Fitzpatrick (1993) and references therein. However, this type of landmark might be very uncomfortable for the patient. In contrast, *soft* landmarks or *retrospective* landmarks are deduced from the images themselves. The spatial location of these "anatomical" landmarks requires expert knowledge and/or sophisticated image analysis tools for automatic detection; see, e.g., Rohr (2001). Other features based on global intensity knowledge will be discussed in Chapter 5.

To make the feature-based registration idea slightly more formal, let $\mathcal{F}(R,j)$ and $\mathcal{F}(T,j)$ denote the jth feature in the reference image R and the template image T, respectively, $j = 1,\ldots,m$, where $m \in \mathbb{N}$ denotes the number of features. The registration problem reads as follows.

Problem 4.1 *Let $m \in \mathbb{N}$ and the features $\mathcal{F}(R,j)$ and $\mathcal{F}(T,j)$, $j = 1,\ldots,m$, be given. Find a transformation $\varphi : \mathbb{R}^d \to \mathbb{R}^d$, such that*

$$\mathcal{F}(R,j) = \varphi(\mathcal{F}(T,j)), \quad j = 1,\ldots,m. \tag{4.1}$$

As we will see subsequently, the interpolation problem 4.1 might also be replaced by the following approximation problem 4.2. For convenience and later usage, we define the *distance measure*

$$\mathcal{D}^{\mathrm{LM}}[\varphi] := \sum_{j=1}^{m} \|\mathcal{F}(R,j) - \varphi(\mathcal{F}(T,j))\|_f^2\,, \tag{4.2}$$

where $\|\cdot\|_f$ denotes a norm on the feature space, e.g., $\|\cdot\|_f = \|\cdot\|_{\mathbb{R}^d}$, if the features are locations of points.

Problem 4.2 *Let* $\mathcal{D}^{\mathrm{LM}}$ *be as in eqn (4.2). Find a transformation* $\varphi : \mathbb{R}^d \to \mathbb{R}^d$*, such that* $\mathcal{D}^{\mathrm{LM}}[\varphi] = \min$.

Suitability means that the transformation is either an element of a typically finite-dimensional space, for example, spanned by polynomials, splines, or wavelets (see Section 3.3.1) or it is required to be smooth in a certain sense to be discussed later. Note that any of these choices may be viewed as a regularization of Problem 4.2.

4.2 Landmark-based parametric registration

In this approach, the features to be mapped are the locations of a number of user-supplied landmarks, i.e., spatial positions,

$$\mathcal{F}(R,j) = x^{R,j}, \quad \mathcal{F}(T,j) = x^{T,j}, \qquad j = 1,\dots,m, \quad m \in \mathbb{N}.$$

Figure 4.1 illustrates the location of a set of six landmarks in a reference image and the corresponding set in a template image.

Using the Euclidean norm in eqn (4.2), we obtain

$$\mathcal{D}^{\mathrm{LM}}[\varphi] = \sum_{j=1}^{m} \left\|x^{R,j} - \varphi(x^{T,j})\right\|_{\mathbb{R}^d}^2. \tag{4.3}$$

For the first approach, we expand the transformation $\varphi = (\varphi_1,\dots,\varphi_d)^\top$ in terms of some basis functions ψ_k, i.e., with some coefficients $\alpha_{\ell,k}$, we have

$$\varphi_\ell = \sum_{k=1}^{n} \alpha_{\ell,k}\psi_k, \qquad \alpha_{\ell,k} \in \mathbb{R}, \quad \psi_k : \mathbb{R}^d \to \mathbb{R}, \quad n \in \mathbb{N}, \quad \ell = 1,\dots,d. \tag{4.4}$$

For simplicity, we assume that the ψ_k's as well as the lengths of the expansions are the same for all $\ell = 1,\dots,d$ and do not depend on ℓ.

With the expansion (4.4), Problem 4.2 can be rewritten as follows.

Problem 4.3 *Find parameters* $\alpha_{\ell,k} \in \mathbb{R}$, $k = 1, \ldots, n$, $\ell = 1, \ldots, d$, *such that* $\mathcal{D}^{\text{LM}}[\varphi] = \min$, *where* $\mathcal{D}^{\text{LM}}$ *is given by eqn (4.3) and* $\varphi = (\varphi_1, \ldots, \varphi_d)^\top$ *is given by eqn (4.4).*

A computation gives

$$\mathcal{D}^{\text{LM}}[\varphi] = \sum_{j=1}^{m} \sum_{\ell=1}^{d} \left(x_\ell^{R,j} - \sum_{k=1}^{n} \alpha_{\ell,k} \psi_k(x^{T,j}) \right)^2 = \sum_{\ell=1}^{d} \| y_\ell - \Psi a_\ell \|_{\mathbb{R}^d}^2 ,$$

where for $\ell = 1, \ldots, d$,

$$\begin{aligned} y_\ell &= (x_\ell^{R,1}, \ldots, x_\ell^{R,m})^\top \in \mathbb{R}^m, \\ \Psi &= \Big(\psi_k(x^{T,j}) \Big)_{\substack{j=1,\ldots,m \\ k=1,\ldots,n}} \in \mathbb{R}^{m \times n}, \\ \text{and} \quad a_\ell &= (\alpha_{\ell,1}, \ldots, \alpha_{\ell,n})^\top \in \mathbb{R}^n. \end{aligned}$$

Thus, the problem of determining the optimal parameters $\alpha_{\ell,k}$ in Problem 4.3 decouples with respect to the spatial dimension ℓ. The optimal parameters can be obtained by solving d least squares problems, where, under the above assumption, the matrix Ψ does not depend on ℓ.

Assuming that Ψ has full rank, $\text{rank}(\Psi) = n$, the solution of Problem 4.3 is unique. A numerical solution can be obtained by using a QR-factorization of Ψ; see, e.g., Golub & van Loan (1989, §5.3.4).

If in particular $\| y_\ell - \Psi a_\ell \|_{\mathbb{R}^d} = 0$ for all $\ell = 1, \ldots, d$, we have a one-to-one correspondence between the landmarks in the reference and template images; see also Fig. 4.1.

For the important case of a linear transformation, we have $n = d + 1$ and

$$\varphi = \begin{pmatrix} \sum_{k=1}^{d+1} \alpha_{1,k} \psi_k \\ \vdots \\ \sum_{k=1}^{d+1} \alpha_{d,k} \psi_k \end{pmatrix} \in \Pi_1^d(\mathbb{R}^d),$$

where $\psi_1(x) = 1$, $\psi_{\ell+1}(x) = x_\ell$, $\ell = 1, \ldots, d$. The restriction $\text{rank}(\Psi) = n = m$ is equivalent to $x^{T,j}$ not being co-linear.

However, the theorem of Mairhuber & Curtis (1956, 1958) (for a modern formulation see, e.g., Braess (1980)) gives a disappointing answer to the question whether an interpolation problem in multi-dimensional spaces has a unique solution or not. The following corollary states that a set of points can be found, such that the Vandermonde matrix becomes singular. Thus, additional conditions are required to make the problem well-posed. For example, using an affine linear interpolation, the interpolation points are not allowed to be co-linear.

Corollary 4.1 *Let $d, m \in \mathbb{N}$, ψ_k, $k = 1, \dots, m$, be continuous functions, $\psi_k : \mathbb{R}^d \to \mathbb{R}$. If $d > 1$ and $m > 1$, then there exists a set of points $x_j \in \mathbb{R}^d$, $j = 1, \dots, m$, such that* $\det(\psi_k(x_j)_{j,k=1,\dots,n}) = 0$.

Proof Assume there exists a set of points $x_j \in \mathbb{R}^d$, $j = 1, \dots, m$, such that $\det(\psi_k(x_j)_{j,k=1,\dots,n}) = c \neq 0$ and hence $x_j \neq x_k$ for $k \neq j$. Since $n \geq 2$, we can find a closed, non-intersecting curve $\gamma : [0, 1] \to \mathbb{R}^d$ with $x_1 = \gamma(0) = \gamma(1)$, $\gamma(1/2) = x_2$, and $x_k \notin \gamma(0, 1])$ for $k = 3, \dots, m$. For the determinant of

$$\Psi(t) := \begin{pmatrix} \psi_1(\gamma(t)) & \psi_2(\gamma(t)) & \cdots & \psi_m(\gamma(t)) \\ \psi_1(\gamma(t + \frac{1}{2})) & \psi_2(\gamma(t + \frac{1}{2})) & \cdots & \psi_m(\gamma(t + \frac{1}{2})) \\ \psi_1(x_3) & \psi_2(x_3) & \cdots & \psi_m(x_3) \\ \vdots & \vdots & \ddots & \vdots \\ \psi_1(x_m) & \psi_2(x_m) & \cdots & \psi_m(x_m) \end{pmatrix} \in \mathbb{R}^{m \times m},$$

we have $\det \Psi(0) = -\det \Psi\left(\frac{1}{2}\right)$, since $\Psi\left(\frac{1}{2}\right)$ can be obtained from $\Psi(0)$ by interchanging the first two rows. Thus there exists a $\xi \in]0, \frac{1}{2}[$, such that $\det \Psi(\xi) = 0$ and thus the Vandermonde matrix with respect to $\gamma(\xi), \gamma(\xi + \frac{1}{2}), x_3, \dots, x_m$ is singular. □

Note that for $d = 1$, the proof does not apply since we cannot find a non-intersecting curve. This shows the particular quality of one-dimensional interpolation.

The main disadvantage is, however, that the transformation from the parametric approach is in general not diffeomorphic. Figure 4.1 shows the results for a linear ($\varphi \in \Pi_1^d(\mathbb{R}^d)$) and a quadratic ($\varphi \in \Pi_2^d(\mathbb{R}^d)$) parametric registration. As is apparent from this figure, the linear approach yields satisfactory results, though the fit of the landmarks is not perfect. After quadratic registration, all landmarks are mapped perfectly. However, since φ is a quadratic polynomial, the map is not diffeomorphic and leads to a “mirrored” image, which is certainly not a satisfactory registration. Note that in this example, the landmarks are chosen such that the interpolation problem is well-posed.

4.3 Landmark-based smooth registration

As already seen in Fig. 4.1, the parametric approach presented in Section 4.2 has some severe drawbacks. Figure 4.2 (LEFT) illustrates these drawbacks for dimension one. The figure shows the results of approximating some monotonic data a linear and a quadratic polynomial. Although the quadratic polynomial

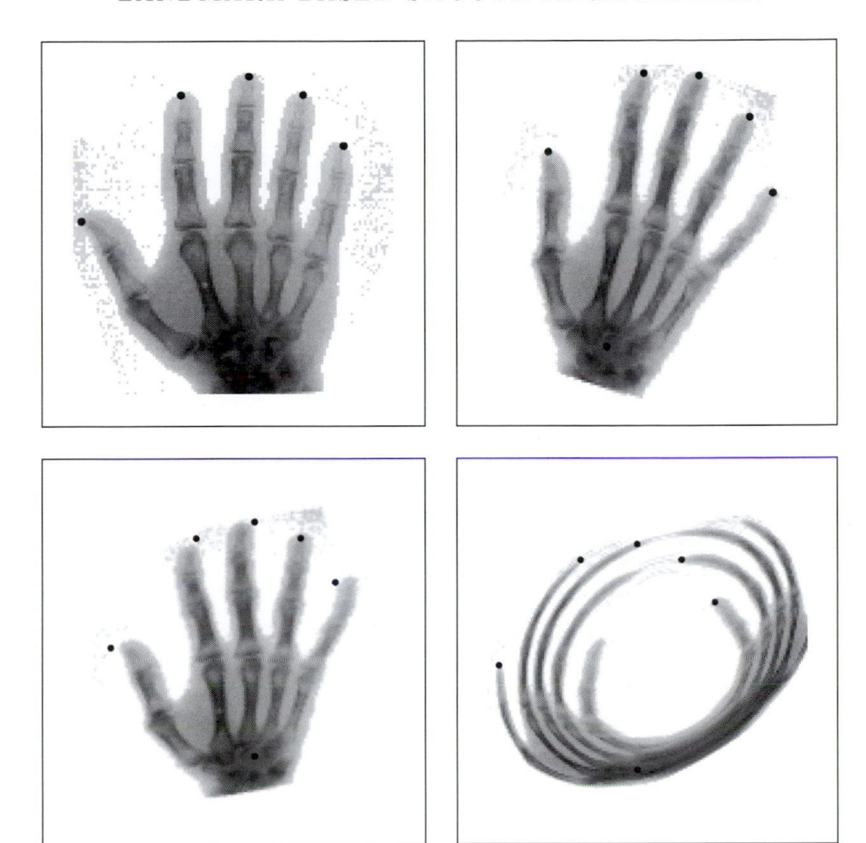

FIG. 4.1 Landmark-based image registration. TOP LEFT: reference, and TOP RIGHT: template with landmarks (black dots), BOTTOM LEFT: parametric linear registered, BOTTOM RIGHT: parametric quadratic registered.

is optimal with respect to the data, it is not preferable for registration. This is because the quadratic is not bijective, manifests oscillation, and does not reflect the monotonicity of the data.

Instead of tuning parameters in an expansion of the transformation in terms of some more or less artificial basis functions, we introduce additional smoothness restrictions to the transformation. These restrictions are expressed by a functional $\mathcal{S}$. Roughly speaking, smoothness is measured in terms of curvature. It turns out, somewhat surprisingly, that the minimizer of this regularized approach is again parameterized: it is a linear combination of shifts of a *radial basis function* plus some polynomial corrections.

In order to provide a detailed insight into the underlying interpolation concepts, we present a general treatment following Light (1995). To begin with, we are looking for an interpolant $\psi : \mathbb{R}^d \to \mathbb{R}$ which is smooth in a certain sense.

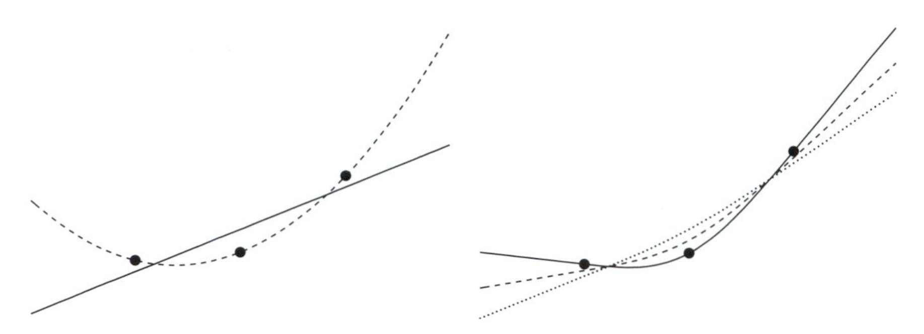

FIG. 4.2 LEFT: approximation with parametric φ, data (dots), linear (solid), and quadratic (dashed) approximations; RIGHT: approximation with non-parametric φ with smoothing parameter $\alpha = 0$ (solid), 1 (dash–dotted), and 5 (dotted).

The smoothest interpolant is called a *minimal norm solution* and it is the interpolant we are interested in. In Section 4.3.1 we show how this interpolant can be derived using the so-called *representers.* As it turns out, the general solution can be expanded in terms of these representers. For polynomial interpolation in a one-dimensional space, this result is well-known and the representers are the Lagrange polynomials.

In Section 4.3.2 we present the basic idea of thin plate spline (TPS) interpolation and in Section 4.3.3 of TPS approximation. TPSs were introduced by the pioneering work of Duchon (1976). In Section 4.3.4 we return to image registration. The performance of TPS-based registration is demonstrated later by Fig. 4.4.

4.3.1 *Minimal norm solutions for interpolation problems*

Following Light (1995) we treat the problem of finding a minimal norm element in a space with interpolation restrictions. To this end, we consider the L_2 inner product

$$\langle f, g \rangle_0 := \langle f, g \rangle_{L_2} := \int_{\mathbb{R}^d} f(x) g(x) dx \tag{4.5}$$

and a semi-inner product of order q,

$$\langle \cdot\,, \cdot \rangle_q : \mathcal{X} \times \mathcal{X} \to \mathbb{R}, \quad \langle f, g \rangle_q := \sum_{|\kappa|=q} c_\kappa \langle D^\kappa f, D^\kappa g \rangle_0 \,, \tag{4.6}$$

where $\kappa \in \mathbb{N}_0^d$, $|\kappa| = \kappa_1 + \cdots + \kappa_d$, $D^\kappa f = \left(\frac{\partial}{\partial x_1}\right)^{\kappa_1} \cdots \left(\frac{\partial}{\partial x_d}\right)^{\kappa_d} f$, $\mathcal{X} = H^q \cap C(\mathbb{R}^d) \cup \Pi_{q-1}(\mathbb{R}^d)$, and H^q denotes the Sobolev space of order q. The set of coefficients $\{ c_\kappa : |\kappa| = q \}$ is chosen such that the semi-norm is rotationally invariant. Explicitly, these parameters are specified via the formal expansion

$$\|x\|_{\mathbb{R}^d}^{2q} = \sum_{|\kappa|=q} c_\kappa x^{2\kappa} = \sum_{|\kappa|=q} c_\kappa x_1^{2\kappa_1} \cdots x_d^{2\kappa_d}. \tag{4.7}$$

In particular for $d = q = 2$, we have

$$\|x\|_{\mathbb{R}^2}^4 = \left(x_1^2 + x_2^2\right)^2 = x_1^4 + 2x_1^2x_2^2 + x_2^4$$

and

$$\langle f, g \rangle_2 = \int_{\mathbb{R}^d} \partial_{x_1x_1} f \, \partial_{x_1x_1} g + 2\partial_{x_1x_2} f \, \partial_{x_1x_2} g + \partial_{x_2x_2} f \, \partial_{x_2x_2} g \; dx.$$

For $q > 0$, eqn (4.6) defines a semi-inner product because it has a non-trivial kernel $K = \{f \in \mathcal{X} : \langle f, f \rangle_q = 0\} = \Pi_{q-1}(\mathbb{R}^d)$.

Suppose a number of interpolation data $(x_j, y_j) \in \mathbb{R}^d \times \mathbb{R}$, $j = 1, \ldots, m$, for $m \in \mathbb{N}$ are given and the following condition holds,

$$[p \in \Pi_{q-1}(\mathbb{R}^d) \wedge p(x_j) = 0, \;\; j = 1, \ldots, m] \Rightarrow p \equiv 0. \tag{4.8}$$

Note that $m \geq d_q := \dim(\Pi_{q-1}(\mathbb{R}^d))$.

Condition (4.8) guarantees that the corresponding Vandermonde matrix has full rank. Moreover,

$$[f, g]_q := \langle f, g \rangle_q + \sum_{j=1}^{m} f(x_j) g(x_j) \tag{4.9}$$

is an inner product on $\mathcal{X}$; cf., e.g., Light (1995).

Our goal is to derive an explicit expression for the *minimal norm solution* ψ, i.e.,

$$\psi = \arg\min\{[f, f]_q, \; f \in H \text{ and } f(x_j) = y_j, \; j = 1, \ldots, m\},$$

where $H := (\mathcal{X}, [\,\cdot\,,\,\cdot\,]_q)$ is a Hilbert space. The minimizer is unique since the norm is convex. In order to compute the minimizer, we construct a representer $R_x \in H$ for $x \in \mathbb{R}^d$, where the representer is characterized by $f(x) = [f, R_x]_q$ for all $f \in H$.

Under the additional assumption

$$\begin{aligned} &\forall x \in \mathbb{R}^d \; \exists K_x \in \mathbb{R}_{\geq 0} : \\ &f(x_1) = \cdots = f(x_m) = 0 \Rightarrow |f(x)| \leq K_x \sqrt{\langle f, f \rangle_q}, \end{aligned} \tag{4.10}$$

the point evaluation functionals (cf., Definition 3.5) are continuous functionals on the Hilbert space $H := (\mathcal{X}, [\,\cdot\,,\,\cdot\,]_q)$. An important fact is that if

$$2q > d,$$

then H consists of continuous functions on $\mathbb{R}^d$, and the point evaluation functionals are elements of the dual H^*; see, e.g., Light (1995).

We now define a subspace $H_0 \subset H$ with integrated interpolation conditions,

$$H_0 := \{ f \in H : f(x_1) = \cdots = f(x_{d_q}) = 0 \},$$

where $x_1, \ldots, x_{d_q}$ are chosen such that condition (4.8) holds for $m = d_q$. We then construct a representer $R_{0,x}$ for a fixed $x \in \mathbb{R}^d$ on the space H_0 and finally extend $R_{0,x}$ to R_x, a representer for x on H.

For $f \notin H_0$, a projection $\mathcal{Q} : H \to H_0$ has to be used. Let $L_1, \ldots, L_{d_q}$ be a Lagrange basis for $\Pi_{q-1}(\mathbb{R}^d)$ and $\mathcal{P} : H \to \Pi_{q-1}(\mathbb{R}^d)$ with

$$\mathcal{P}f := \sum_{j=1}^{d_q} f(x_j) L_j.$$

Hence, $\mathcal{Q} = \mathcal{I} - \mathcal{P}$ maps H onto H_0.

For $u, v \in H$ we define the inner product

$$[u, v]_q = \langle u, v \rangle_q + \sum_{j=1}^{d_q} u(x_j) v(x_j).$$

Note that $[u, f]_q = \langle u, f \rangle_q$ for all $u \in H$ and $f \in H_0$.

For a characterization of $R_{0,x}$, we assume that for a fixed $x \in \mathbb{R}^d$ the function $R_{0,x}$ is a representer on H_0. Thus, we have

$$\begin{aligned} f(x) - \mathcal{P}f(x) &= [f - \mathcal{P}f, R_{0,x}]_q = \langle f - \mathcal{P}f, R_{0,x} \rangle_q = \langle f, R_{0,x} \rangle_q \\ &= \sum_{|\kappa|=q} c_\kappa \int_{\mathbb{R}^d} (D^\kappa f)(y)(D^\kappa R_{0,x})(y) dy \\ &= \sum_{|\kappa|=q} c_\kappa \langle D^\kappa f, D^\kappa R_{0,x} \rangle_0 \\ &= \left\langle f, (-1)^q \sum_{|\kappa|=q} c_\kappa D^{2\kappa} R_{0,x} \right\rangle_0 , \end{aligned} \tag{4.11}$$

where Green's formula (see, e.g., Buck (1978, §9.4)) has been used. On the other hand

$$f(x) - \mathcal{P}f(x) = f(x) - \sum_{j=1}^{d_q} f(x_j) L_j(x) = \left\langle f, \delta_x - \sum_{j=1}^{d_q} L_j(x) \delta_{x_j} \right\rangle_0 . \tag{4.12}$$

Equations (4.11) and (4.12) show that $R_{0,x}$ is the solution of the distributional differential equation

$$(-1)^q \sum_{|\kappa|=q} c_\kappa D^{2\kappa} R_{0,x} = \delta_x - \sum_{j=1}^{d_q} L_j(x)\delta_{x_j}. \tag{4.13}$$

Note that the right hand side of eqn (4.13) is a linear combination of point evaluation functionals. Thus, the solution can be derived from combinations of shifted versions of a fundamental solution or radial basis function.

Let $\rho_x(y) := \rho(\|y-x\|_{\mathbb{R}^d})$, where ρ is the *radial basis function* or Green's function; see Theorem 4.2 and Rohr (2001).

Theorem 4.2 *The radial basis function for* $(-1)^q \sum_{|\kappa|=q} c_\kappa D^{2\kappa}$ *is given by*

$$\rho(r) := c_q^d \begin{cases} r^{2q-d}\log r, & d \text{ even}, \\ r^{2q-d}, & d \text{ odd}, \end{cases} \tag{4.14}$$

where

$$c_q^d = \begin{cases} \frac{(-1)^{q+1+d/2}}{2^{2q-1}\pi^{d/2}(q-1)!(q-d/2)!}, & d \text{ even}, \\ \frac{\Gamma(d/2-q)}{2^{2q}\pi^{d/2}(q-1)!}, & d \text{ odd}. \end{cases}$$

With this radial basis function we have

$$(-1)^q \sum_{|\kappa|=q} c_\kappa D^{2\kappa}\rho_x = \delta_x$$

and thus a particular solution of eqn (4.13) is given by

$$\widetilde{R}_{0,x} = \rho_x - \sum_{j=1}^{d_q} L_j(x)\rho_{x_j}.$$

Since $\widetilde{R}_{0,x}$ is not necessarily in H_0 we define

$$\begin{aligned} R_{0,x} &:= (\mathcal{I}-\mathcal{P})\widetilde{R}_{0,x} \\ &= \rho_x - \sum_{j=1}^{d_q} L_j(x)\rho_{x_j} - \sum_{j=1}^{d_q} \rho_x(x_j)L_j - \sum_{j,k=1}^{d_q} \rho_{x_j}(x_k)L_j(x)L_k. \end{aligned} \tag{4.15}$$

It remains to extend $R_{0,x}$ to the full space H. To this end we make use of the following lemma.

Lemma 4.3 (Light (1995, Th. 2.1)) *Let x_j, $j = 1, \ldots, d_q$, be distinct points in $\mathbb{R}^d$ satisfying condition (4.8) for $m = d_q = \dim(\Pi_{q-1}(\mathbb{R}^d))$. Then the Lagrange polynomial L_k is a representer for x_k in H, $k = 1, \ldots, d_q$.*

Proof For any $f \in H$ we have

$$[f, L_k]_q = \langle f, L_k \rangle_q + \sum_{j=1}^{d_q} f(x_j) L_k(x_j) = \langle f, L_k \rangle_q + f(x_k) = f(x_k),$$

since $L_k \in \Pi_{q-1}(\mathbb{R}^d)$ belongs to the kernel of the semi-inner product. □

Now we are in a position to construct the representer R_x in H. For $f \in H$ we have

$$\begin{aligned} [f, R_{0,x}]_q &= [f - \mathcal{P}f, R_{0,x}]_q + [\mathcal{P}f, R_{0,x}]_q = f(x) - \mathcal{P}f(x) + 0 \\ &= f(x) - \sum_{j=1}^{d_q} f(x_j) L_j(x) = f(x) - \sum_{j=1}^{d_q} [f, L_j]_q L_j(x) \end{aligned}$$

or

$$f(x) = [f, R_{0,x}]_q + \sum_{j=1}^{d_q} [f, L_j]_q L_j(x) = \left[f, R_{0,x} + \sum_{j=1}^{d_q} L_j(x) L_j \right]_q,$$

showing that

$$R_x := R_{0,x} + \sum_{j=1}^{d_q} L_j(x) L_j$$

is a representer in H.

The following theorem gives an explicit expansion of the minimal norm solution in terms of representers.

Theorem 4.4 (Light (1995, Th. 1.1)) *Let $f \in H$ with $f(x_j) = y_j$, $j = 1, \ldots, m$, and let $\psi \in C_f$ be the minimal norm solution, where*

$$C_f := \left\{ v \in H : \ [v, v]_q \leq [f, f]_q \wedge v(x_j) = y_j, \ j = 1, \ldots, m \right\}.$$

Let R_x be the representer for $x \in \mathbb{R}^d$. Then there exist $\theta_1, \ldots, \theta_m \in \mathbb{R}$, such that $\psi = \sum_{j=1}^m \theta_j R_{x_j}$. The coefficients $\theta_1, \ldots, \theta_m$ are determined by the equations

$$y_k = \psi(x_k) = [\psi, R_{x_k}]_q = \sum_{j=1}^{m} \theta_j \left[R_{x_j}, R_{x_k} \right]_q.$$

However, a characterization explicitly based on coefficients in an expansion with respect to basis functions is of particular interest. The next Theorem 4.5 characterizes the minimal norm solution with respect to the radial basis functions.

Theorem 4.5 *Let $d, m, q \in \mathbb{N}$ and $d_q := \dim(\Pi_{q-1}(\mathbb{R}^d))$ and let $x_j \in \mathbb{R}^d$, $y_j \in \mathbb{R}$, $j = 1, \ldots, m$, be given interpolation data. The minimal norm solution*

$$\psi = \arg\min\{[f, f]_q,\ f \in H \text{ and } f(x_j) = y_j,\ j = 1, \ldots, m\}$$

is characterized by

$$\psi = \sum_{j=1}^{m} \theta_j \rho_{x_j} + \sum_{j=1}^{d_q} \beta_j p_j, \tag{4.16}$$

where $\rho_{x_j} = \rho(\|\cdot - x_j\|_{\mathbb{R}^d})$ (see Theorem 4.2) and $p_1, \ldots, p_{d_q}$ is a basis for $\Pi_{q-1}(\mathbb{R}^d)$. The coefficients $\theta := (\theta_1, \ldots, \theta_m)^\top \in \mathbb{R}^m$ and $\beta := (\beta_1, \ldots, \beta_{d_q})^\top \in \mathbb{R}^{d_q}$ are determined by the following system of linear equations,

$$\begin{pmatrix} K & B^\top \\ B & 0 \end{pmatrix} \begin{pmatrix} \theta \\ \beta \end{pmatrix} = \begin{pmatrix} y \\ 0 \end{pmatrix}, \tag{4.17}$$

$$y := (y_1, \ldots, y_m)^\top \in \mathbb{R}^m,$$

$$K := (\rho(\|x_j - x_k\|_{\mathbb{R}^d}))_{j,k=1,\ldots,m} \in \mathbb{R}^{m \times m}, \tag{4.18}$$

$$\text{and} \quad B := (p_j(x_k))_{\substack{j=1,\ldots,d_q \\ k=1,\ldots,m}} \in \mathbb{R}^{d_q \times m}. \tag{4.19}$$

Proof In the expansion $\psi = \sum_{j=1}^{m} \mu_j R_{x_j}$ (see Theorem 4.4) any representer is a linear combination of translated radial basis functions and polynomials. Thus, with some coefficients θ_j and β_j together with a basis $p_1, \ldots, p_{d_q}$ for $\Pi_{q-1}(\mathbb{R}^d)$ we have the characterization given by eqn (4.16), which satisfies the equations

$$y_k = \psi(x_k) = \sum_{j=1}^{m} \theta_j \rho_{x_j}(x_k) + \sum_{j=1}^{d_q} \beta_j p_j(x_k), \quad k = 1, \ldots, m, \tag{4.20}$$

or $K\theta + B^\top \beta = y$.

Moreover, since $\Pi_{q-1}(\mathbb{R}^d)$ is the kernel of $\langle \cdot\,, \cdot \rangle_q$, we have for any $p \in \Pi_{q-1}(\mathbb{R}^d)$,

$$\begin{aligned} 0 = \langle \psi, p \rangle_q &= \sum_{|\kappa|=q} c_\kappa \int_{\mathbb{R}^d} (D^\kappa \psi)(D^\kappa p)\, dx \\ &= \sum_{j=1}^m \theta_j \int_{\mathbb{R}^d} (-1)^q \sum_{|\kappa|=q} c_\kappa (D^{2\kappa} \rho_{x_j}) p \, dx \\ &= \sum_{j=1}^m \theta_j \left\langle (-1)^q \sum_{|\kappa|=q} c_\kappa D^{2\kappa} \rho_{x_j}, p \right\rangle_0 \\ &= \sum_{j=1}^m \theta_j \left\langle \delta_{x_j}, p \right\rangle_0 = \sum_{j=1}^m \theta_j p(x_j). \end{aligned} \tag{4.21}$$

Expanding p with respect to $p_1, \ldots, p_{d_q}$, we also find $B\theta = 0$. □

For a modest number of interpolation data, the system of linear equations (4.17) can be solved using standard techniques. Some results are shown in Section 4.3.4, where we also compare this approach with one based on approximation. For a higher number of interpolation data this is still a topic of research; cf., e.g., Schaback (1997).

4.3.2 *Example: splines and TPSs*

Now we study the important cases $d = 1 \wedge q = 2$ and $d = q = 2$ in detail. With the semi-inner product (4.6) and the abbreviation

$$\mathcal{S}^{\text{TPS}}[\psi] := \tfrac{1}{2} \langle \psi, \psi \rangle_q^2 \tag{4.22}$$

we consider the following problem.

Problem 4.4 *Find $\psi : \mathbb{R}^d \to \mathbb{R}$, such that*

$$\mathcal{S}^{\text{TPS}}[\psi] = \min \;\; \textit{subject to } \psi(x_j) = y_j, \; j = 1, \ldots, m.$$

We will derive an explicit solution for this problem. The Lagrange function for Problem 4.4 reads

$$L[\psi, \lambda] = \mathcal{S}^{\text{TPS}}[\psi] + \sum_{j=1}^m \lambda_j (\delta_{x_j}[\psi] - y_j),$$

where λ_j are Lagrange multipliers and δ_z are the point evaluation functionals (cf., Definition 3.5) located at position z, i.e.,

$$\delta_z[\psi] := \int_{\mathbb{R}^d} \delta_z(x)\psi(x)dx = \psi(z).$$

We start by computing the Gâteaux derivative of δ_z,

$$\begin{aligned} d\delta_z[\psi;\zeta] &:= \lim_{h\to 0}\frac{1}{h}(\delta_z[\psi+h\zeta]-\delta_z[\psi]) \\ &= \lim_{h\to 0}\frac{1}{h}(\psi(z)+h\zeta(z)-\psi(z)) = \delta_z[\zeta], \end{aligned}$$

and the Gâteaux derivative of $\mathcal{S}^{\mathrm{TPS}}$,

$$\begin{aligned} d\mathcal{S}^{\mathrm{TPS}}[\psi;\zeta] &:= \lim_{h\to 0}\frac{1}{h}(\mathcal{S}^{\mathrm{TPS}}[\psi+h\zeta]-\mathcal{S}^{\mathrm{TPS}}[\psi]) \\ &= \sum_{|c_\kappa|=q} c_\kappa \lim_{h\to 0}\frac{1}{2h}\int_{\mathbb{R}^d}(D^\kappa[\psi+h\zeta])^2-(D^\kappa\psi)^2dx \\ &= \sum_{|c_\kappa|=q} c_\kappa \int_{\mathbb{R}^d}(D^\kappa\psi)\,(D^\kappa\zeta)dx \\ &= (-1)^q\sum_{|c_\kappa|=q} c_\kappa \int_{\mathbb{R}^d}(D^{2\kappa}\psi)\,\zeta dx, \end{aligned}$$

for suitable perturbations ζ. Here, we used Green's formula (cf., e.g., Buck (1978, §9.4)) and, as an implicit necessary condition for a minimizer, $\psi(x)\to 0$ for $\|x\|_{\mathbb{R}^d}\to\infty$.

For $q=2$, we have

$$(-1)^q\sum_{|c_\kappa|=q} c_\kappa D^{2\kappa}\psi = \Delta^2\psi = \begin{cases} \psi^{(\mathrm{IV})}, & d=1 \\ \partial_{1111}\psi + 2\partial_{1122}\psi + \partial_{2222}\psi, & d=2; \end{cases}$$

cf., eqn (4.7). A minimizer for Problem 4.4 must satisfy the condition

$$dL[\psi,\lambda;\zeta] = \int_{\mathbb{R}^d}\left(\Delta^2\psi(x) + \sum_{j=1}^m \lambda_j\delta_{x_j}(x)\right)\zeta(x)dx = 0$$

for all perturbations ζ. Hence by variation of ζ, we obtain the distributional differential equation

$$\Delta^2\psi(x) + \sum_{j=1}^m \lambda_j\delta_{x_j} = 0. \tag{4.23}$$

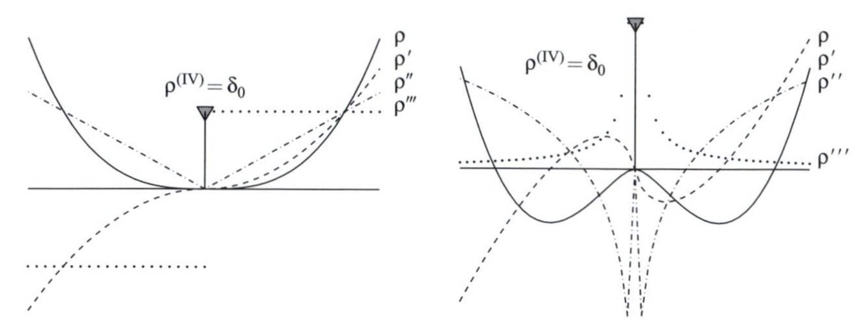

FIG. 4.3 Fundamental solution ρ and derivatives for $d = 1$ (LEFT) and $d = 2$ (RIGHT).

A fundamental solution of $\Delta^2\psi(x) + \delta_0 = 0$ is the Green's function ψ_0, with

$$\psi_0(x) = \rho(\|x\|_{\mathbb{R}^d}) + p_3(x), \quad p_3 \in \Pi_3(\mathbb{R}^d),$$

where with $r := \|x\|_{\mathbb{R}^d}$,

$$\rho(r) = \begin{cases} \frac{1}{12}r^3, & d = 1 \\ \frac{1}{8\pi}r^2 \log r, & d = 2; \end{cases}$$

see also Fig. 4.3.

For $d = 1$ this is verified straighforwardly; for $d = 2$ and $x \neq 0$,

$$\begin{aligned} \nabla\rho(x) &= (8\pi)^{-1}(2\log r + 1)x, \\ \Delta\rho(x) &= (2\pi)^{-1}(\log r + 1), \\ \nabla(\Delta\rho)(x) &= (2\pi r^2)^{-1}x, \\ \Delta^2\rho(x) &= (\pi r^3)^{-1}\langle\nabla r, x\rangle + (2\pi r^2)^{-1}\operatorname{div} x = 0. \end{aligned}$$

Thus the general solution of Problem 4.4 takes the form

$$\psi = \sum_{j=1}^{m} \theta_j \rho_{x_j} + \sum_{j=1}^{3} \beta_j p_j, \quad \operatorname{span}\{p_1, p_2, p_3\} = \Pi_1(\mathbb{R}^d),$$

$$\psi(x_j) = y_j, \quad \text{for } j = 1, \ldots, m,$$

which is, of course, in accordance with Theorem 4.4 and Theorem 4.5.

For the case $d = 1$ we rediscovered the well-known interpolating *cubic splines*, and the solution for $d = 2$ is also known as *thin plate splines*; cf., e.g., Rohr (2001). Note that for the one-dimensional case there exists an $\mathcal{O}(m)$ algorithm based on moments (second order derivatives of the spline in the interpolation points); cf., e.g., Piegl & Tiller (1997, §2.6).

4.3.3 *Smooth minimal norm approximations*

We now slightly extend the theory derived in Section 4.3.1. To this end, we consider the inner product

$$[u, v]_{q,\alpha} := \alpha \langle u, v \rangle_q + \sum_{j=1}^{m} (u(x_j) - y_j)(v(x_j) - y_j), \tag{4.24}$$

where $\alpha > 0$ serves as a regularizing parameter. A necessary condition for a minimal norm solution ψ now reads

$$\alpha(-1)^q \sum_{|\kappa|=q} c_\kappa D^{2\kappa}\psi + \sum_{j=1}^{m} \left(\langle \delta_{x_j}, \psi \rangle_0 - y_j\right) \delta_{x_j} = 0.$$

Exploiting the expansion

$$\psi = \sum_{j=1}^{m} \theta_j \rho_{x_j} + \sum_{j=1}^{d_q} \beta_j p_j \tag{4.25}$$

(cf., Theorem 4.5) we have

$$\begin{aligned} 0 &= \alpha(-1)^q \sum_{|\kappa|=q} c_\kappa D^{2\kappa}\psi + \sum_{j=1}^{m} \left(\langle \delta_{x_j}, \psi \rangle_0 - y_j\right) \delta_{x_j} \\ &= \alpha \sum_{j=1}^{m} \theta_j \left((-1)^q \sum_{|\kappa|=q} c_\kappa D^{2\kappa} \rho_{x_j} \right) + \sum_{j=1}^{m} \left(\langle \delta_{x_j}, \psi \rangle_0 - y_j\right) \delta_{x_j} \\ &= \sum_{j=1}^{m} \left(\alpha\theta_j + \left(\langle \delta_{x_j}, \psi \rangle_0 - y_j\right)\right) \delta_{x_j}. \end{aligned}$$

Hence, for $k = 1, \ldots, m$,

$$0 = \alpha\theta_k + \left(\langle \delta_{x_k}, \psi \rangle_0 - y_k\right)$$

or, equivalently, $\psi(x_k) + \alpha\theta_k = y_k$. This shows that the optimal parameters θ and β in the expansion (4.25) are determined by

$$\begin{pmatrix} K + \alpha I_m & B^\top \\ B & 0 \end{pmatrix} \begin{pmatrix} \theta \\ \beta \end{pmatrix} = \begin{pmatrix} y \\ 0 \end{pmatrix}, \tag{4.26}$$

where K and B are as in Theorem 4.5 and I_m denotes the identity matrix. Interpolation thus turns out to be a special case of approximation if we choose $\alpha = 0$ in the above equation.

Figure 4.2 (RIGHT) demonstrates the role of the regularization parameter α. For $\alpha = 0$, the function interpolates the data. Varying $\alpha = 1, 5$, we found that the function ψ becomes more and more linear (less curved).

4.3.4 *Smooth transformations*

We now have the ingredients to solve the registration problems 4.5 and 4.6, respectively. We again make use of the abbreviation

$$\mathcal{S}^{\mathrm{TPS}}[\varphi] := \frac{1}{2}\sum_{\ell=1}^{d} \langle \varphi_\ell, \varphi_\ell \rangle_q ,$$

where $\langle \cdot , \cdot \rangle_q$ is the semi-inner product defined by eqn (4.6) with kernel $\Pi_{q-1}(\mathbb{R}^d)$.

Let $x^{T,j}$ and $x^{R,j}$, $j = 1,\ldots,m$, be the given landmarks in the reference and template image, respectively, and let ρ denote the radial basis function with respect to the semi-inner product. Moreover, we define the interpolation space

$$\mathrm{IS}_{q,m} := \mathrm{span}\{\rho(\|x - x^{T,j}\|_{\mathbb{R}^d}),\ j = 1,\ldots,m\}^d \cup \Pi^d_{q-1}(\mathbb{R}^d). \tag{4.27}$$

Problem 4.5 *Find a transformation $\varphi : \mathbb{R}^d \to \mathbb{R}^d$, such that*

$$\mathcal{S}^{\mathrm{TPS}}[\varphi] \longrightarrow \min$$

subject to $\varphi(x^{T,j}) = x^{R,j}$, $j = 1,\ldots,m$.

Problem 4.6 *Given $\alpha > 0$, find a transformation $\varphi : \mathbb{R}^d \to \mathbb{R}^d$, $\varphi \in \mathrm{IS}_{q,m}$, such that*

$$\alpha\mathcal{S}^{\mathrm{TPS}}[\varphi] + \mathcal{D}^{\mathrm{LM}}[\varphi] \longrightarrow \min,$$

where $\mathcal{D}^{\mathrm{LM}}$ is defined by eqn (4.3) and $\mathrm{IS}_{q,m}$ by eqn (4.27).

The solution of both problems has the form

$$\varphi_\ell = \sum_{j=1}^{m} \theta_{\ell,j}\rho_{x_j} + \sum_{j=1}^{d_q} \beta_{\ell,j} p_j, \quad \ell = 1,\ldots,d,$$

where ρ is the radial basis function corresponding to the semi-inner product, $d_q := \dim(\Pi^d_{q-1})$, and $p_1,\ldots,p_{d_q}$ is a basis of Π^d_{q-1}. For $\ell = 1,\ldots,d$, the coefficients $\theta_\ell := (\theta_{\ell,1},\ldots,\theta_{\ell,m})^\top \in \mathbb{R}^m$ and $\beta_\ell := (\beta_{\ell,1},\ldots,\beta_{\ell,d_q})^\top \in \mathbb{R}^{d_q}$ are determined by

$$\begin{pmatrix} K + \alpha I & B^\top \\ B & 0 \end{pmatrix} \begin{pmatrix} \theta_\ell \\ \beta_\ell \end{pmatrix} = \begin{pmatrix} y_\ell \\ 0 \end{pmatrix}, \tag{4.28}$$

where $y_\ell := (x_\ell^{R,1},\ldots,x_\ell^{R,m})^\top \in \mathbb{R}^m$ and the matrices K and B are defined in Theorem 4.5. Note that K and B do not depend on ℓ.

Again, the solution of Problem 4.5 might be viewed as a particular solution of Problem 4.6 for $\alpha = 0$.

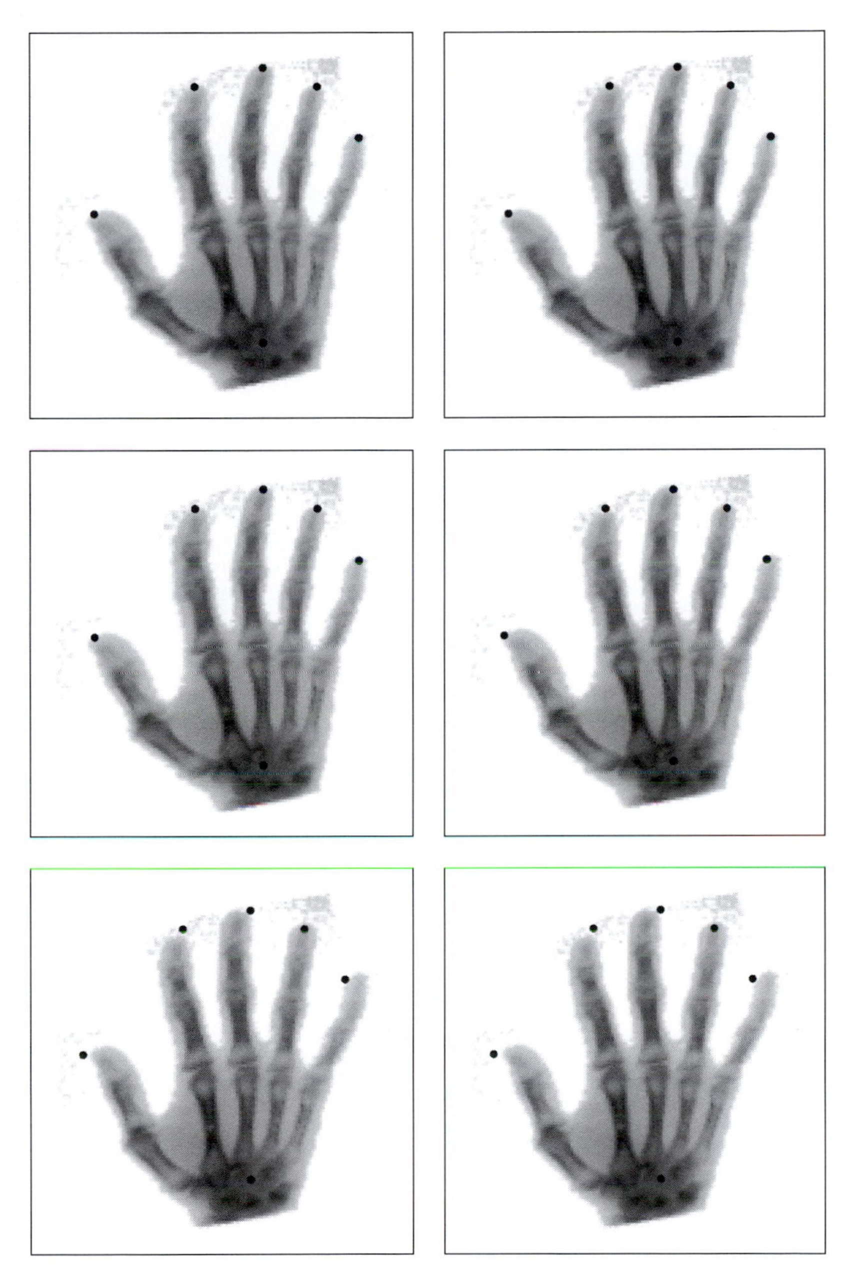

FIG. 4.4 TPS-based registration for $\alpha = 0, 10, 10^2, 10^3, 10^4,$ and 10^5 and (LEFT to RIGHT and TOP to BOTTOM) respectively.

4.4 An example of landmark-based registration

Figure 4.4 displays results for the landmark-based registration of reference and template images which have already been shown in Fig. 4.1. As is apparent from this figure, the registration ranges from an interpolation of the landmarks ($\alpha = 0$) to almost affine linear registration for large values of α. Note that the registration is governed completely by the landmarks.

Finally, Fig. 4.5 illustrates that landmark-based registration does not always result in a meaningful registration. The landmarks are chosen such that one expects a bending of the rectangular bar displayed in the template image. Note that the landmarks are chosen in a meaningful ordering. However, although the transformation is smooth, it fails to be diffeomorphic. This can be seen in the bottom right picture, where the sign of the Jacobian of the transformation, i.e., $\mathrm{sign}(\det \nabla\varphi)$, is shown.

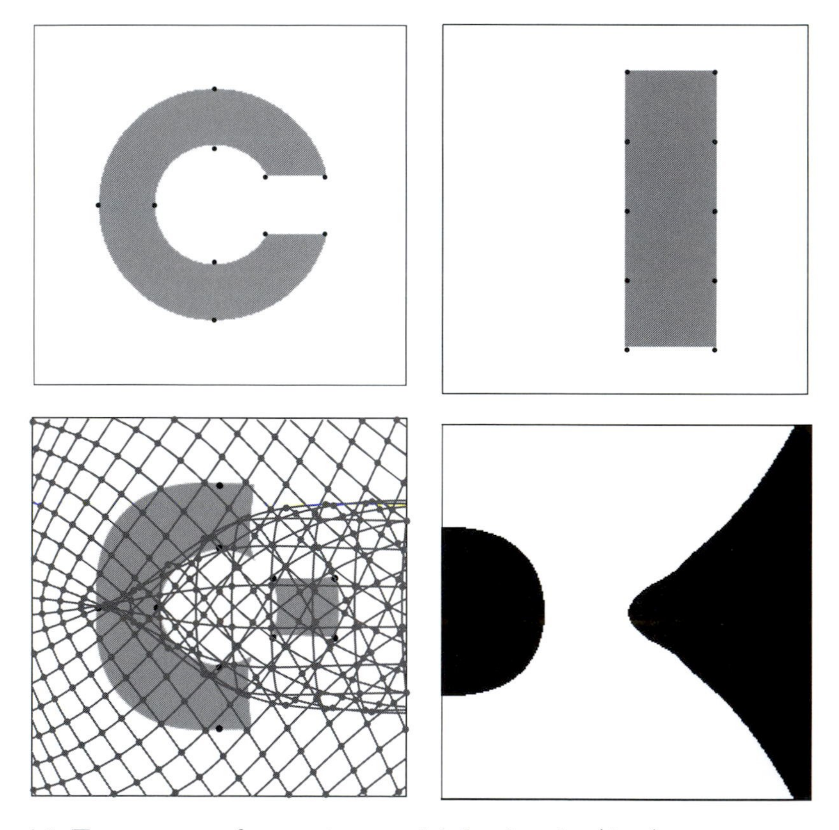

FIG. 4.5 TOP LEFT: reference image with landmarks (dots), TOP RIGHT: template image with landmarks (dots), BOTTOM LEFT: template image after TPS registration with landmarks and deformed grid, BOTTOM RIGHT: sign of determinant of Jacobian of φ (white: $\det \nabla\varphi \geq 0$, black: $\det \nabla\varphi < 0$).

5

PRINCIPAL AXES-BASED REGISTRATION

The registration technique introduced in Chapter 4 has a major drawback. That is, the registration process is governed by the location and correspondence of the landmarks.

Although there are many sophisticated ideas for automatically locating the landmarks (see, e.g., Rohr (2001) and references therein) the process is still not fully automated. Moreover, the location of landmarks might be very complicated, since even experts are not always able to characterize mathematically, for example, anatomical landmarks in medical images.

In this chapter we follow the idea of registering image features, but now the registration is based on features which can be deduced from the images automatically. To this end, we consider an image as a density function or mass distribution. From this distribution we derive the so-called *principal axes*. The registration is also called *principal axes transformation* (PAT). For the relevant literature, we refer to Maurer & Fitzpatrick (1993) and references therein, and particularly Alpert et al (1990).

Definition 5.1 *Let* $d \in \mathbb{N}$ *and* $B : \mathbb{R}^d \to \mathbb{R}$ *be an image; cf., Definition 3.1. We define the* expectation value *of a function* f *with respect to* B *by*

$$\mathbb{E}_B[f] := \frac{\int_{\mathbb{R}^d} f(x)\, B(x)dx}{\int_{\mathbb{R}^d} B(x)dx}.$$

For $u : \mathbb{R}^d \to \mathbb{R}^{m\times n}$, *we set* $\mathbb{E}_B[u] := (\mathbb{E}_B[u_{j,k}])_{\substack{j=1,\dots,m\\k=1,\dots,n}} \in \mathbb{R}^{m\times n}$.

The center *of an image is defined by* $c_B := \mathbb{E}_B[x] \in \mathbb{R}^d$ *and the* covariance *by* $\mathrm{Cov}_B := \mathbb{E}_B\left[(x-c_B)(x-c_B)^\top\right] \in \mathbb{R}^{d\times d}$.

The center and an eigendecomposition of the covariance matrix are used as features for the image. The resulting registration technique is named principal axes transformation (PAT); see, e.g., Alpert et al (1990).

Since the covariance matrix is real, symmetric, and positive semi-definite, it permits an orthogonal eigendecomposition

$$\mathrm{Cov}_B = D_B \Sigma_B^2 D_B^\top, \tag{5.1}$$

where $D_B \in \mathbb{R}^{d\times d}$ is unitary, i.e., $D_B^\top D_B = I_d$, and the matrix

$$\Sigma_B = \operatorname{diag}(\sigma_{B,1}, \ldots, \sigma_{B,d}) \in \mathbb{R}^{d\times d}$$

is diagonal; cf., e.g., Golub & van Loan (1989). For normalization purposes, we arrange the columns of D_B such that for the *standard deviations* we have $\sigma_{B,1} \geq \cdots \geq \sigma_{B,d} \geq 0$. If the eigenvalues of Cov_B are simple, the decomposition is essentially unique, up to a sign in the columns of D_B.

5.1 Geometrical interpretation

Suppose we are looking for an optimal representation of the mass distribution B by a straight line $G = p + \mathbb{R}y$ (the *principal axis*), where $\|y\|_{\mathbb{R}^d} = 1$. The distance of an arbitrary point $x \in \mathbb{R}^d$ to this line is given by

$$d_{p,y}(x) := \|x - p - \langle x - p, y\rangle_{\mathbb{R}^d}\, y\|_{\mathbb{R}^d}$$

and the optimal line can be derived from the minimization of

$$D(p, y) \longrightarrow \min,$$

where

$$D(p, y) = \mathbb{E}_B\left[(d_{p,y}(x))^2\right] = \mathbb{E}_B\left[\langle x - p, x - p\rangle_{\mathbb{R}^d} - \langle x - p, y\rangle_{\mathbb{R}^d}^2\right].$$

Necessary conditions for a minimizer (p^*, y^*) are

$$\begin{aligned} 0 &= \nabla_p D(p^*, y^*) \\ &= \mathbb{E}_B\left[-2(x - p^*) + 2\langle x - p^*, y^*\rangle_{\mathbb{R}^d}\, y^*\right] \\ &= -2(c_B - p^*) + 2\langle c_B - p^*, y^*\rangle_{\mathbb{R}^d}\, y^* \end{aligned}$$

and

$$\begin{aligned} 0 &= \nabla_y D(p^*, y^*) \\ &= \mathbb{E}_B\left[-2\langle x - p^*, y^*\rangle_{\mathbb{R}^d}\, (x - p^*)\right] \\ &= -2\langle c_B - p^*, y^*\rangle_{\mathbb{R}^d}\, (c_B - p^*). \end{aligned}$$

These equations show that $p^* = c_B$ is optimal. Thus,

$$D(p^*, y) = \mathbb{E}_B\left[(x - c_B)^\top (x - c_B)\right] - y^\top \mathrm{Cov}_B y,$$

and the minimum is attained for y^* being any eigenvector of length one belonging to the eigenspace of the largest eigenvalue of the covariance matrix Cov_B. In other words, a line through the center in the direction of an eigenvector corresponding to the largest eigenvalue, a principal axis, is an optimal representation of the image.

5.2 Stochastic interpretation

Suppose we are looking for a Gaussian density

$$g_{\Sigma,\mu}(x) := (2\pi)^{-d/2} \det(\Sigma)^{-1} \exp\left(-\tfrac{1}{2}(x-\mu)^\top (\Sigma\Sigma^\top)^{-1}(x-\mu)\right)$$

which fits the density given by the image B optimally in the sense that the so-called log-*likelihood* is maximized,

$$\mathbb{E}_B\left[\log g_{\Sigma,\mu}\right] \xrightarrow{\Sigma,\mu} \max .$$

Elementary computations give

$$\begin{aligned}
&\mathbb{E}_B\left[\log g_{\Sigma,\mu}\right] \\
&\quad = -\frac{d}{2}\log(2\pi) - \log(\det\Sigma) - \mathbb{E}_B\left[\frac{1}{2}(x-\mu)^\top(\Sigma\Sigma^\top)^{-1}(x-\mu)\right].
\end{aligned}$$

Differentiation with respect to μ shows that $\mu = c_B = \mathbb{E}_B[x]$ is optimal. Now,

$$\begin{aligned}
&2\mathbb{E}_B\left[\log g_{\Sigma,c_B}\right] + d\log(2\pi) \\
&\quad = 2\log(\det(\Sigma^{-1})) - \mathbb{E}_B\left[\operatorname{trace}\left(\Sigma^{-1}(x-c_B)(x-c_B)^\top\Sigma^{-\top}\right)\right] \\
&\quad = -\log(\det \mathrm{Cov}_B) + \log(\det(\Sigma^{-1}\mathrm{Cov}_B\Sigma^{-\top})) - \operatorname{trace}(\Sigma^{-1}\mathrm{Cov}_B\Sigma^{-\top}).
\end{aligned}$$

Since $A := \Sigma^{-1}\mathrm{Cov}_B\Sigma^{-\top}$ is real, symmetric, and positive semi-definite, it permits an orthogonal eigendecomposition. Hence, the maximization of the log-likelihood is equivalent to the maximization of

$$\log(\det(A)) - \operatorname{trace}(A) = \sum_{j=1}^{d}(\log\lambda_j - \lambda_j),$$

where the eigenvalues of A are denoted by $\lambda_1, \ldots, \lambda_d$. This finally shows that the log-likelihood attains its maximum if and only if $\mu = c_B$ and $\lambda_j = 1$ for $j = 1, \ldots, d$. Hence, $I_d = A = \Sigma^{-1}\mathrm{Cov}_B\Sigma^{-\top}$ or $\mathrm{Cov}_B = \Sigma\Sigma^\top$.

Summarizing, the best possible description of the image B in the class of Gaussian densities is given by

$$g_B := (2\pi)^{-d/2}(\det \mathrm{Cov}_B)^{-1/2} \exp\left(-\tfrac{1}{2}(x-c_B)^\top \mathrm{Cov}_B^{-1}(x-c_B)\right).$$

Features of the reference density g_B can be used for a description of the image B.

Note that with $m := \int_{\mathbb{R}^d} B(x)dx$, we have

$$\begin{aligned} m\mathbb{E}_B[\log f] &= \int_{\mathbb{R}^d} \log f(x)B(x)\,dx \\ &= \int_{\mathbb{R}^d} \left[\log \frac{f(x)}{B(x)} + \log B(x)\right] B(x)\,dx \\ &= \int_{\mathbb{R}^d} \log \frac{f(x)}{B(x)} B(x)dx + \int_{\mathbb{R}^d} \log B(x)B(x)\,dx, \end{aligned}$$

where the first term is the non-symmetric so-called Kullback–Leibler distance between f and B (cf., Kullback & Leibler (1951)) and the second term is the negative entropy of the image B. Thus, the previous approach may also be interpreted as a minimization of the Kullback–Leibler distance too.

5.3 A robust generalization

The main advantage of this second interpretation given in Section 5.2 is that it can be generalized. Since the Gaussian distribution (referred to as *standard*) is not robust with respect to perturbations, one might wish to replace it for example by the Cauchy distribution (referred to as *robust*).

Kent & Tyler (1988) showed that the Gaussian density might be replaced by the density of the t-distribution on $\nu > 0$ degrees of freedom. The density of the t-distribution is given by

$$g_{\Sigma,\mu}(x) = C_{\nu,p} \det \Sigma^{-1}(1 + \nu^{-1}\|\Sigma^{-1}(x-\mu)\|^2)^{-(\nu+d)/2},$$

where $C_{\nu,p}$ is some suitable normalization constant which is not dependent on Σ or μ; see, e.g., Mardia et al (1979). In contrast to the Gaussian distribution, which can be viewed as a special t-distribution for $\nu \to \infty$, where the optimal parameter can be computed explicitly, the solution for a general t-distribution is only known in terms of a fixed-point type of equation

$$M = \mathbb{E}_1\left[u_\nu(x^\top M^{-1}x)\, xx^\top\right] \tag{5.2}$$

where, with $\lambda > 0, M$ is an augmented moment matrix,

$$M = \begin{pmatrix} \Sigma\Sigma^\top + \lambda^{-1}\mu\mu^\top & \lambda^{-1}\mu \\ \lambda^{-1}\mu^\top & \lambda^{-1} \end{pmatrix} \in \mathbb{R}^{(d+1)\times(d+1)},$$

and $u_\nu(t) = (\nu + d)/(\nu + t)$; see Kent & Tyler (1991).

Equation (5.2) can be solved numerically using a fixed-point iteration as shown in Kent & Tyler (1991).

5.4 A simple comparison with $d = 2$

The performance of the standard and robust approaches is illustrated for the spatial dimension $d = 2$. We take advantage of an eigendecomposition of the covariance matrix Cov_B,

$$\mathrm{Cov}_B = D(\rho_B)\Sigma_B^2 D(-\rho_B),$$

where for $d = 2$,

$$D(\rho_B) = \begin{pmatrix} \cos\rho_B & -\sin\rho_B \\ \sin\rho_B & \cos\rho_B \end{pmatrix}, \quad \Sigma_B = \begin{pmatrix} \sigma_{B,1} & 0 \\ 0 & \sigma_{B,2} \end{pmatrix};$$

cf., eqn (5.1). The direction of the principal axis is given by $(\cos\rho_B, \sin\rho_B)^\top$ and the direction of the secondary axis is given by $(-\sin\rho_B, \cos\rho_B)^\top$. Thus, the angle ρ_B can be used to describe both axes.

Figure 5.1 illustrates these quantities. The figure shows the principal (solid) and secondary axis (dashed), where the lengths of the axes indicate the standard deviations. The center $c_B = (c_{B,1}, c_{B,2})^\top$ is the intersection point of the two axes. These five quantities might by viewed as an image feature, i.e.,

$$\mathcal{F}_B := (c_{B,1}, c_{B,2}, \sigma_{B,1}, \sigma_{B,2}, \rho_B). \tag{5.3}$$

The four pictures in Fig. 5.1 illustrate this feature for different PAT approaches for an unperturbed image:

1. *standard approach* based on the Gaussian density (cf., Section 5.2),
2. *robust approach* based on the Cauchy density (cf., Section 5.3),
3. *standard approach* for the binary image, and
4. *robust approach* for the binary image.

In Fig. 5.2 we show analogous results for the artificially perturbed image $\hat{B}$. The numerical values for these examples are summarized in Table 5.1.

As is apparent from this example, the standard approach (based on a Gaussian density) is much more sensitive with respect to a perturbation of the data than the robust approach (based on a Cauchy density). The impact of the perturbation can be dampened using the binary image instead of the image itself as a reference density. However, ignoring the intensity might not be a good idea in many applications since typically the intensity is meaningful.

5.5 Principal axes under transformation

We are interested in the transformation properties of the feature (5.3) under an affine linear map. In other words, we are interested in the center and covariance matrix of the image $\hat{B} := B \circ \varphi$ where $\varphi : \mathbb{R}^d \to \mathbb{R}^d$ is affine linear, $\varphi(x) = Ax + b$, $\det A > 0$.

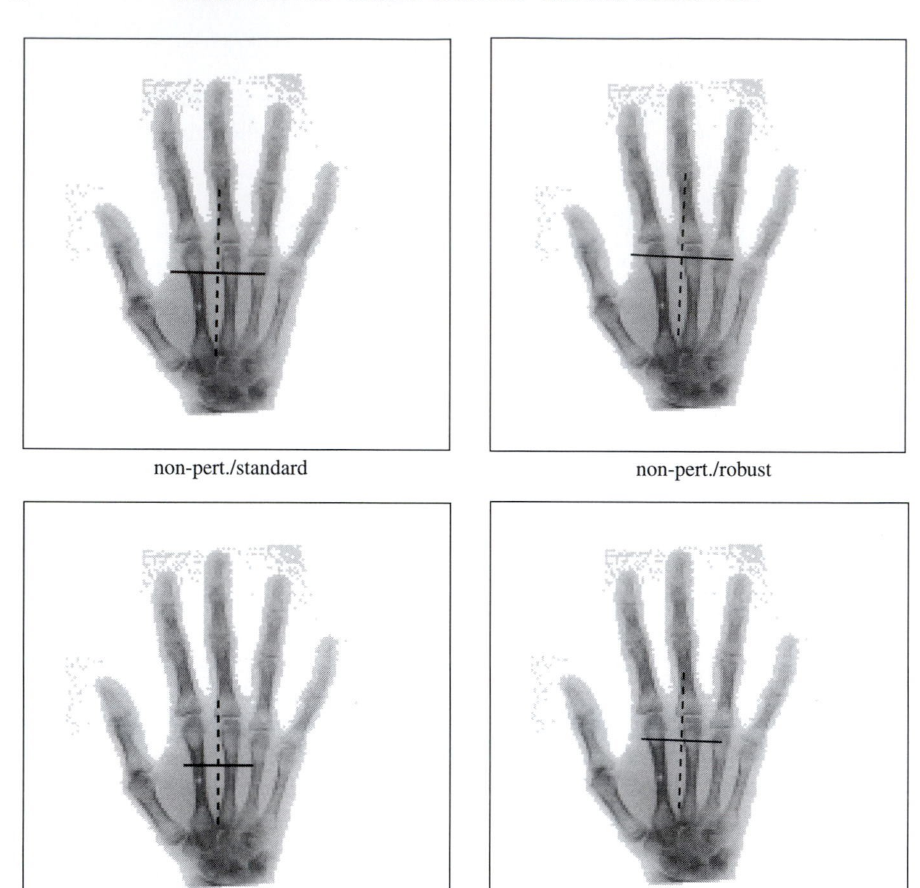

FIG. 5.1 Non-perturbed images of a human hand with principal axis (solid) and secondary axis (dashed); the intersection point indicates the location of the center c_B and the length of an axis illustrates the corresponding eigenvalue of Cov_B.

Theorem 5.1 *Let $B : \mathbb{R}^d \to \mathbb{R}$ be an image with center c_B and covariance matrix Cov_B; cf., Definition 5.1. Moreover, let $\hat{B}(x) := B(Ax + b)$, where $A \in \mathbb{R}^{d \times d}$, $\det(A) > 0$, and $b \in \mathbb{R}^d$. Then*

$$c_B = Ac_{\hat{B}} + b \quad \text{and} \quad \mathrm{Cov}_B = A\mathrm{Cov}_{\hat{B}}A^\top.$$

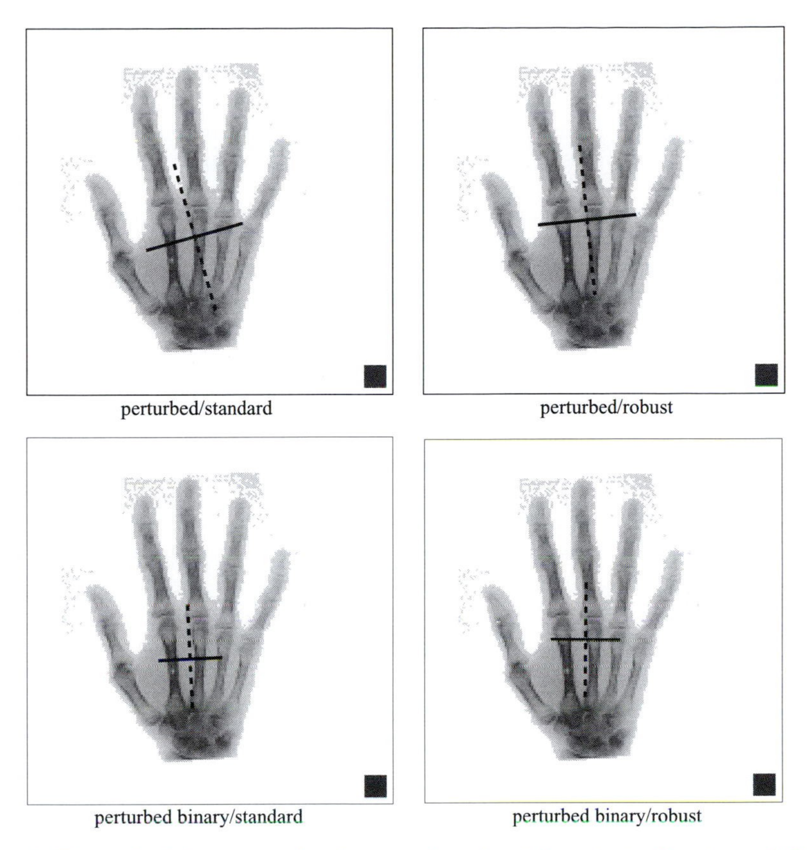

FIG. 5.2 Perturbed images of a human hand with principal axis (solid) and secondary axis (dashed); the intersection point indicates the location of the center c_B and the length of an axis illustrates the corresponding eigenvalue of Cov_B.

Proof From the transformation theorem (cf., e.g., Buck (1978, Theorem 6, §8.3)) with $y = \varphi(x) = Ax + b$ and $dy = \det A\, dx$, we have

$$\int_{\mathbb{R}^d} f(y)B(y)dy = \det(A) \int_{\mathbb{R}^d} f \circ \varphi(x)\hat{B}(x)dx$$

and hence,

$$\begin{aligned}
\mathbb{E}_B\left[f\right] &= \mathbb{E}_{\hat{B}}\left[f \circ \varphi\right], \\
c_B &= \mathbb{E}_B\left[x\right] = A\mathbb{E}_{\hat{B}}\left[x\right] + b = Ac_{\hat{B}} + b, \\
\mathrm{Cov}_B &= \mathbb{E}_B\left[(x - c_B)(x - c_B)^\top\right] \\
&= \mathbb{E}_{\hat{B}}\left[A(x - c_{\hat{B}})(x - c_{\hat{B}})^\top A^\top\right] = A\mathrm{Cov}_{\hat{B}}A^\top.
\end{aligned}$$

□

Table 5.1 *Statistics of the standard and robust approaches for an image of a human hand (original image B), an artificially perturbed image $\hat{B}$, and the binary images of B and $\hat{B}$. The differences in the stochastic features are shown as well. The computations are performed on 128^2 pixel images.*

	Standard			Robust		
	B	$\hat{B}$	diff.	B	$\hat{B}$	diff.
$c_{B,1}$	91.64	93.85	2.2120	98.57	98.85	0.2829
$c_{B,2}$	72.95	75.81	2.8594	73.15	73.52	0.3711
$\sigma_{B,1}$	17.96	22.24	4.2765	13.21	14.04	0.8371
$\sigma_{B,2}$	31.32	33.60	2.2753	24.18	24.24	0.0509
ρ_B	−1.59	−1.30	0.2909	−1.57	−1.53	0.0448
	bin(B)	bin($\hat{B}$)	diff.	bin(B)	bin($\hat{B}$)	diff.
$c_{B,1}$	85.65	86.78	1.1259	90.18	90.36	0.1847
$c_{B,2}$	73.46	74.78	1.3223	73.86	74.08	0.2180
$\sigma_{B,1}$	19.84	22.11	2.2741	15.31	15.70	0.3949
$\sigma_{B,2}$	32.30	33.30	1.0034	25.52	25.57	0.0554
ρ_B	−1.62	−1.47	0.1459	−1.61	−1.58	0.0266

Theorem 5.1 is a starting point for registration purposes. The idea is to carry out some normalization of the images. Exploiting two maps φ_R and φ_T, the reference image R and the template image T are mapped to $\widetilde{R}$ and $\widetilde{T}$, respectively, such that $c_{\widetilde{R}} = c_{\widetilde{T}} = 0$ and $\mathrm{Cov}_{\widetilde{R}} = \mathrm{Cov}_{\widetilde{T}} = I_d$. Hence, $\varphi = \varphi_T \varphi_R^{-1}$ gives a registration of T. This idea is the topic of the next theorem.

Theorem 5.2 *Let R and T be two images with centers c_R and c_T and non-singular covariance matrices Cov_R and Cov_T, respectively. Defining $\hat{T}(x) := T(Ax + b)$, where*

$$A := D_T \Sigma_T U \Sigma_R^{-1} D_R^\top \in \mathbb{R}^{d\times d}, \quad b := c_T - Ac_R \in \mathbb{R}^d,$$

$U \in \mathbb{R}^{d\times d}$ is an arbitrary unitary matrix, and the eigendecompositions $\mathrm{Cov}_R = D_R \Sigma_R^2 D_R^\top$ and $\mathrm{Cov}_T = D_T \Sigma_T^2 D_T^\top$ are used.

Then we have

$$c_{\hat{T}} = c_R \quad \textit{and} \quad \mathrm{Cov}_{\hat{T}} = \mathrm{Cov}_R.$$

Proof Note that A is non-singular and well-defined since Cov_R and Cov_T are non-singular. Using Theorem 5.1 we obtain

$$\begin{aligned} c_{\hat{T}} &= A^{-1}(c_T - b) = c_R, \\ \mathrm{Cov}_{\hat{T}} &= A^{-1}\mathrm{Cov}_T(A^{-1})^\top \\ &= [D_R \Sigma_R U^\top \Sigma_T^{-1} D_T^\top][D_T \Sigma_T^2 D_T^\top][D_T \Sigma_T^{-1} U \Sigma_R D_R^\top] \\ &= D_R \Sigma_R U^\top U \Sigma_R D_R^\top \\ &= \mathrm{Cov}_R. \end{aligned}$$

□

The computation of the transformation in Theorem 5.2 involves $d(d+1)$ parameters, the entries of b and A. However, because of the symmetry of the covariance matrices, only $d/2(d+3)$ parameters are determined by Theorem 5.1. Thus, for example, for dimension $d = 2, 3, 4$ there is a gap of one, three, and six equations; see also Schormann & Zilles (1997). In the formulation of Theorem 5.2, this ambiguity is expressed by the unitary matrix U. Here, we find an additional $d/2(d-1)$ degrees of freedom.

Additional ambiguities enter into play, if the covariance has multiple eigenvalues. If the images exhibit a d-dimensional ball, the covariance is a multiple of the matrix I_d. Hence, no particular direction can be selected. However, this situation almost never occurs for the images in real-life applications.

5.6 An example of principal axis-based registration

Figure 5.3 demonstrates that the PAT can be used for image registration successfully. However, Theorem 5.2 already indicates that there is ambiguity in the choice of the unitary matrix U. This ambiguity is illustrated in Fig. 5.4, where

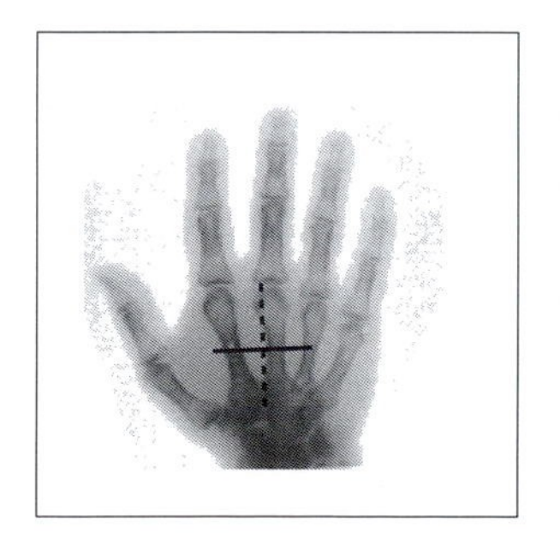
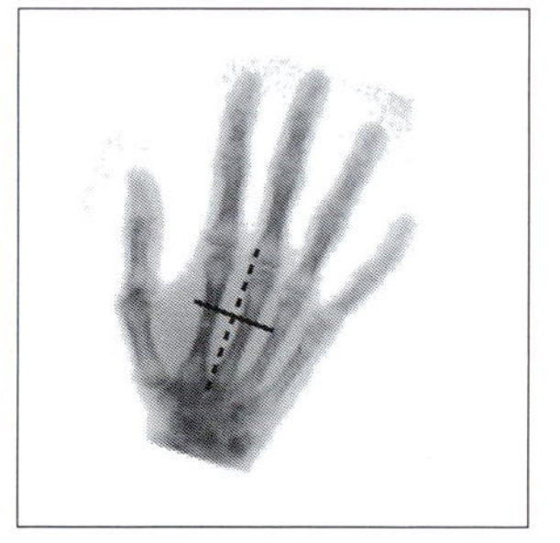
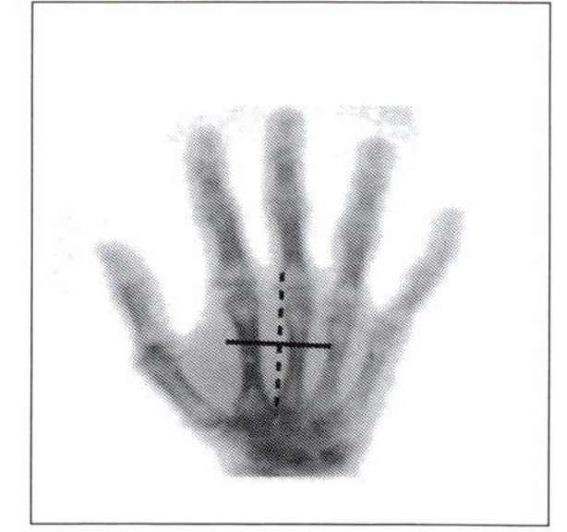

FIG. 5.3 PAT registration images from hands using the robust approach. Reference image R (LEFT), template image T (MIDDLE), and transformed template $\hat{T}$ (RIGHT).

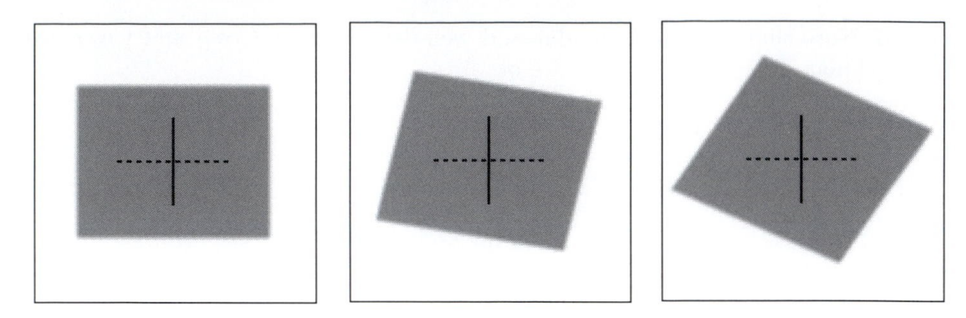

FIG. 5.4 Three different rectangles sharing the same stochastic features.

three different rectangles are depicted. Although these rectangles look quite different, they all share the same stochastic features. As a consequence, the PAT cannot distinguish between these different images.

6

OPTIMAL LINEAR REGISTRATION

In this chapter we investigate the question of how to find an optimal linear registration based on a distance measure $\mathcal{D}$. An analytical solution cannot be expected for the images from our application. Thus, we have to look for a numerical solution. Moreover, since the images under consideration are of high resolution, we focus on fast and efficient schemes, which typically exploit derivatives. Thus, an important property of the distance measure to be discussed is its differentiability.

The choice of an appropriate distance measure is a difficult task. Popular choices to be discussed in the subsequent sections are based on *intensity* (see, e.g., Brown (1992)), *correlation* (see, e.g., Collins & Evans (1997)), or *mutual information* (see, e.g., Viola (1995) or Collignon et al (1995)).

For some particular applications, modifications of these similarity measures have been investigated; see, e.g., Studholme et al (1996) or Roche et al (1999). In addition, the distance measure used in the registration may also be based on particular image features, e.g., edges or surfaces.

We start by introducing a set of feasible transformations, which here are supposed to be affine linear maps, i.e., $\varphi \in \Pi_1^d(\mathbb{R}^d)$; cf., Definition 3.6. A mathematical formulation of the registration problem then reads as follows.

Problem 6.1 *Find* $\varphi \in \Pi_1^d(\mathbb{R}^d)$ *such that* $\mathcal{D}[\varphi] = \min$.

The essential point here is that the set $\in \Pi_1^d(\mathbb{R}^d)$ can be parameterized. For a specific element φ of $\in \Pi_1^d(\mathbb{R}^d)$, we make use of the notation φ_a, where

$$\varphi_{a;\ell}(x) = a_{\ell,0} + \sum_{j=1}^{d} a_{\ell,j} x_j, \quad \ell = 1, \ldots, d.$$

The parameters $a_{\ell,j}$ are gathered together in a vector,

$$a = (a_{1,0}, \ldots, a_{1,d}, \ldots, \quad a_{d,0}, \ldots, a_{d,d})^\top \in \mathbb{R}^n, \quad n = d(d+1).$$

Moreover, we set

$$D(a) := \mathcal{D}[\varphi_a] \quad \text{and} \quad T_a := T \circ \varphi_a. \tag{6.1}$$

Thus, Problem 6.1 may be reformulated in terms of a parameterized finite-dimensional optimization problem.

Problem 6.2 *Find $a \in \mathbb{R}^n$, such that $D(a) = \min$.*

Since differentiability is a major point for the development of fast numerical schemes, we require the template image to be differentiable.

In principle any minimization technique can be used for the minimization of D. However, on the one hand it turns out that *direct methods* or *steepest descent* methods are not fast enough, while on the other hand, second order derivative-based Newton-type methods are not stable for real-life applications. This is because the derivatives of the images have to be approximated from the discrete data. Since these data are typically corrupted by noise, estimating a derivative becomes a delicate matter.

6.1 Intensity-based registration

A straightforward approach is based on the minimization of the so-called *sum of squared differences* (SSD); cf., e.g., Brown (1992) or Čapek (1999).

Definition 6.1 *Let $d \in \mathbb{N}$ and $R, T \in \mathrm{Img}(d)$. The* sum of squared differences (SSD) *distance measure $\mathcal{D}^{\mathrm{SSD}}$ is defined by $\mathcal{D}^{\mathrm{SSD}} : \mathrm{Img}(d)^2 \to \mathbb{R}$,*

$$\mathcal{D}^{\mathrm{SSD}}[R,T] := \frac{1}{2}\|T - R\|^2_{L_2} = \frac{1}{2}\int_{\mathbb{R}^d} \big(T(x) - R(x)\big)^2 dx.$$

For a transformation $\varphi : \mathbb{R}^d \to \mathbb{R}^d$ we also define

$$\mathcal{D}^{\mathrm{SSD}}[R,T;\varphi] = \mathcal{D}^{\mathrm{SSD}}[R, T \circ \varphi], \tag{6.2}$$

and for a parametric transformation φ_a we set

$$D^{\mathrm{SSD}}(R,T;a) = \mathcal{D}^{\mathrm{SSD}}[R, T \circ \varphi_a]. \tag{6.3}$$

In order to make Newton-type methods applicable, we compute the derivatives of D; cf., eqn (6.3). Elementary computations give

$$\partial_{a_j} D^{\mathrm{SSD}}(R,T;a) = \big\langle (T_a - R), \partial_{a_j} T_a \big\rangle_{L_2},$$

$$\partial_{a_j a_k} D^{\mathrm{SSD}}(R,T;a) = \big\langle \partial_{a_j} T_a, \partial_{a_j} T_a \big\rangle_{L_2} + \big\langle (T_a - R), \partial_{a_j, a_k} T_a \big\rangle_{L_2}.$$

Note that we could also work with subderivatives if the images are non-smooth. However, in all our applications, the images are given in terms of discrete data, and we obtain the continuous images by using an interpolation scheme. It is this interpolated image which has to fulfill the smoothness constraints. Thus, the smoothness restriction is less severe as it might appear at first right.

As already pointed out, a second order derivative-based approach might not be the method of choice. Here, we take advantage of the so-called Gauss–Newton method, where, roughly speaking, the linearization step is performed within the norm. However, Levenberg–Marquardt techniques are also used in the literature; cf., e.g., Thévenaz et al (1998).

A first order Taylor expansion gives

$$\begin{aligned} D^{\mathrm{SSD}}(R,T;a+b) &= \tfrac{1}{2}\|T_{a+b} - R\|_{L_2}^2 \\ &\approx \tfrac{1}{2}\|T_a - R + \nabla_a T_a^\top b\|_{L_2}^2. \end{aligned} \tag{6.4}$$

Minimizing the term on the right hand side with respect to b, we discover a linear least squares problem. Thus, for a fixed a, the optimal solution is characterized by the *normal equations*

$$M(a)b = f(a),$$

where $M(a) := (m_{j,k}(a)) \in \mathbb{R}^{n\times n}$, $f(a) = (f_j(a)) \in \mathbb{R}^n$, with

$$f_j(a) = \left\langle T_a - R, \partial_{a_j} T_a \right\rangle_{L_2}, \tag{6.5}$$

$$m_{j,k}(a) = \left\langle \partial_{a_j} T_a, \partial_{a_k} T_a) \right\rangle_{L_2}. \tag{6.6}$$

The overall algorithm for computing an optimal parameter a^* is summarized in Algorithm 6.1. The computationally expensive parts are the $d(d+3)/2$ inner products for the computation of $f(a)$ and $M(a)$ and the computation of $T \circ \varphi_a$. Here, an interpolation scheme has to be exploited. From a theoretical point of view, this approach requires at least a quadratic interpolation scheme. However,

Algorithm 6.1 *Gauss–Newton method for the minimization of D^{SSD}; cf., eqn (6.3).*

Set $k = 0$, choose initial guess $a^{(k)}$.
While not STOP,
  compute $f(a^{(k)})$ and $M(a^{(k)})$,
  cf., eqns (6.5) and (6.6), respectively;
  solve the system of linear equations
  $M(a^{(k)})b = f(a^{(k)})$;
  update $a^{(k+1)} = a^{(k)} + b$, $k \mapsto k+1$;
end.

The iteration is stopped once the norm of the update b is brought below a certain tolerance (here, $\mathrm{tol}_1 = 10^{-5}$) or the relative decrease in the objective function is too slow, $f(a^{(k)}) - f(a^{(k+1)}) \leq \mathrm{tol}_2 f(a^{(k+1)})$. In our implementation we used $\mathrm{tol}_2 = 10^{-6}$.

numerical experience provides evidence that d-linear interpolation schemes as introduced in Section 3.1.3 can be used with success, too.

In the case of discrete images, it is more stable to replace the normal equations by the solution of the finite-dimensional least squares problem using a QR-decomposition; cf., e.g., Golub & van Loan (1989, §5.3.4).

From a mathematical point of view, the type of parameterization of the transformation φ is of no particular importance. Thus, any of the restricted models introduced in Section 3.3.1 can be treated in a similar fashion.

6.2 An example of intensity-based affine linear registration

The results of an optimal intensity-based affine linear registration are shown in Fig. 6.1. Here, the intensity values of the template image have been modified and the minimization is performed with respect to both the geometrical parameters

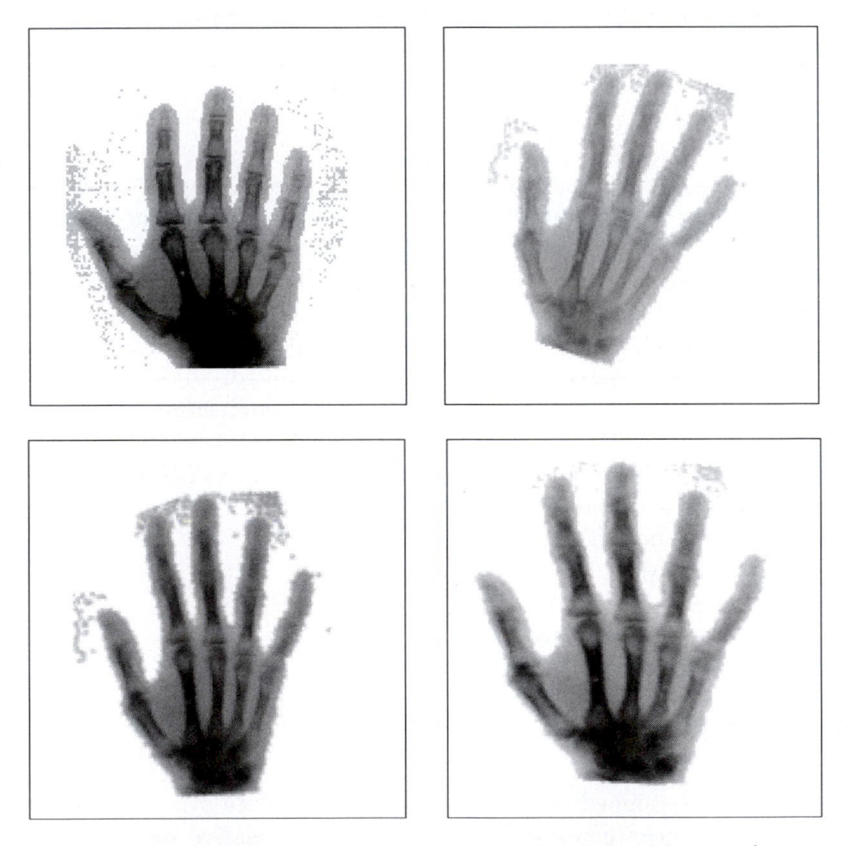

FIG. 6.1 Optimal intensity-based affine linear registration; reference (TOP LEFT), template (TOP RIGHT), template after rigid registration (BOTTOM LEFT), template after affine linear registration (BOTTOM RIGHT).

and the parameters of an affine linear model for the gray values. Two registration results are shown, one for a rigid registration and one for a general affine linear map.

6.3 Correlation-based registration

Registrations based on modifications of the so-called *correlation* have been studied by various authors; see, e.g., Collins & Evans (1997). Here we use the definition introduced by Gonzales & Woods (1993, p. 583).

Definition 6.2 *Let $d \in \mathbb{N}$ and $R, T \in \text{Img}(d)$. The* correlation *between R and T is given by*

$$\text{Corr} : \text{Img}(d)^2 \times \mathbb{R}^d \to \mathbb{R}, \quad \text{Corr}_{R,T}(y) := \int_{\mathbb{R}^d} R(x)T(x-y)dx.$$

The correlation may also be viewed as the L_2-inner product between R and $T(\cdot - y)$. If, in particular, R and T are normalized, such that they are of unit length, the correlation is the cosine of the angle between the two images. Maximization of the correlation with respect to y gives an image $T(\cdot - y)$ which is close to R in the sense that R and $T(\cdot - y)$ are maximally linearly dependent.

The usual normalization (see, e.g., Gonzales & Woods (1993, p. 583) is by statistics of the first kind. To this end, we assume that the support of all images under consideration is contained in a region $\Omega \subset \mathbb{R}^d$. For simplicity and without loss of generality we assume $\Omega =]0,1[^d$ and thus $|\Omega| := \int_\Omega dx = 1$.

Definition 6.3 *Let $d \in \mathbb{N}$ and $B \in \text{Img}(d)$ be an image. The* expectation value *μ and the* standard deviation *σ of B are defined by*

$$\mu(B) := |\Omega|^{-1} \int_\Omega B(x)dx \quad \text{and} \quad \sigma(B) := \mu\Big((B - \mu(B))^2\Big).$$

Definition 6.4 *Let $d \in \mathbb{N}$. The* correlation coefficient *is defined by $\gamma : \text{Img}(d)^2 \times \mathbb{R}^d \to \mathbb{R}$,*

$$\gamma(R, T; y) := \left\langle \frac{R - \mu(R)}{\sigma(R)}, \frac{T_y - \mu(T_y)}{\sigma(T_y)} \right\rangle_{L_2},$$

where $T_y(x) = T(x-y)$ and $\mu(B)$ and $\sigma(B)$ are defined in Definition 6.3.

Using this normalization the correlation coefficient is just the cosine of the angle between $R - \mu(R)$ and $T_y - \mu(T_y)$.

Definition 6.5 *Let* $d \in \mathbb{N}$ *and* $R, T \in \mathrm{Img}(d)$. *The* correlation-*based distance measure* $\mathcal{D}^{\mathrm{corr}}$ *is defined by* $\mathcal{D}^{\mathrm{corr}} : \mathrm{Img}(d)^2 \to \mathbb{R}$,

$$\mathcal{D}^{\mathrm{corr}}[R,T] := \left\langle \frac{R - \mu(R)}{\sigma(R)}, \frac{T - \mu(T)}{\sigma(T)} \right\rangle_{L_2},$$

where μ *and* σ *are defined by Definition 6.3. For a transformation* $\varphi : \mathbb{R}^d \to \mathbb{R}^d$ *we also define*

$$\mathcal{D}^{\mathrm{corr}}[R,T;\varphi] = \mathcal{D}^{\mathrm{corr}}[R, T \circ \varphi], \tag{6.7}$$

and for a parametric transformation φ_a *we set*

$$D^{\mathrm{corr}}(R,T;a) = \mathcal{D}^{\mathrm{corr}}[R, T \circ \varphi_a]. \tag{6.8}$$

Since

$$2\mathcal{D}^{\mathrm{SSD}}[R,T] = \|R\|_{L_2}^2 + \|T\|_{L_2}^2 - 2\langle R,T\rangle_{L_2}$$

and

$$\begin{aligned} \sigma(R)\sigma(T) \cdot \mathcal{D}^{\mathrm{corr}}[R,T] &= \langle R - \mu(R), T - \mu(T)\rangle_{L_2} \\ &= \langle R,T\rangle_{L_2} - \mu(R)\mu(T) \end{aligned}$$

we see that there is a strong connection between the minimization of $\mathcal{D}^{\mathrm{SSD}}$ and the maximization of $\mathcal{D}^{\mathrm{corr}}$. If in particular the transformation is restricted to pure translation, we have $\det(\nabla\varphi) = 1$ and the approaches coincide. This follows from $\|R\|_{L_2}, \sigma(R)$, and $\mu(R)$ and $\|T\|_{L_2}, \sigma(T)$, and $\mu(T)$ being constant, respectively.

6.4 Mutual information-based registration

Since 1995, *mutual information* has been used in image registration. This approach was proposed independently by Viola (1995) and Collignon et al (1995) and has been used since then by many authors, e.g., Kim et al (1997), Maes et al (1997), Gaens et al (1998), Meihe et al (1999), or Abram (2000).

6.4.1 *Mutual information*

The basic idea is the maximization of the so-called *mutual information* of the images with respect to the transformation. Mutual information is an entropy-based measure with a widespread use in information theory. The precise definitions of the distance measure $\mathcal{D}$, the mutual information MI, and the entropy H are summarized as follows.

Definition 6.6 *Let $q \in \mathbb{N}$ and ρ be a density on $\mathbb{R}^q$, i.e., $\rho : \mathbb{R}^q \to \mathbb{R}$, $\rho(x) \geq 0$, and $\int_{\mathbb{R}^q} \rho(x)\, dx = 1$. The (differential)* entropy *of the density is defined by*

$$H(\rho) := -\mathbb{E}_\rho\left[\log \rho\right] = -\int_{\mathbb{R}^q} \rho \log \rho \; dg.$$

Definition 6.7 *Let $d \in \mathbb{N}, R, T \in \mathrm{Img}(d)$. The* mutual information (MI) *distance measure $\mathcal{D}^{\mathrm{MI}}$ is defined by $\mathcal{D}^{\mathrm{MI}} : \mathrm{Img}(d)^2 \to \mathbb{R}$,*

$$\mathcal{D}^{\mathrm{MI}}[R, T] := H(\rho_R) + H(\rho_T) - H(\rho_{R,T}),$$

where ρ_R, ρ_T, and $\rho_{R,T}$ denote the gray-value densities of R, T, and the joint gray-value distribution, respectively.

For a transformation $\varphi : \mathbb{R}^d \to \mathbb{R}^d$ we also define

$$\mathcal{D}^{\mathrm{MI}}[R, T; \varphi] = \mathcal{D}^{\mathrm{MI}}[R, T \circ \varphi], \tag{6.9}$$

and for a parametric transformation φ_a we set

$$D^{\mathrm{MI}}(R, T; a) = \mathcal{D}^{\mathrm{MI}}[R, T \circ \varphi_a]. \tag{6.10}$$

The basic idea of mutual information is illustrated by Fig. 6.2. Here, the transformed templates $T \circ \varphi$, where φ is essentially a rotation of degree α, and the joint gray value density $\rho_{T,T\circ\varphi}$ are displayed. This figure shows that the density is very "sharp", when $T_\varphi = T$ and becomes "smeared out" when α increases. Since the mutual information essentially measures the entropy of the joint density, it is maximal if the images are maximally related.

The entropy is the expectation of the negative logarithm of the density. Thus we may also write

$$\mathcal{D}^{\mathrm{MI}}[R, T] = -\mathbb{E}_{\rho_{R,T}}\left[\log \frac{\rho_{R,T}}{\rho_R \rho_T}\right],$$

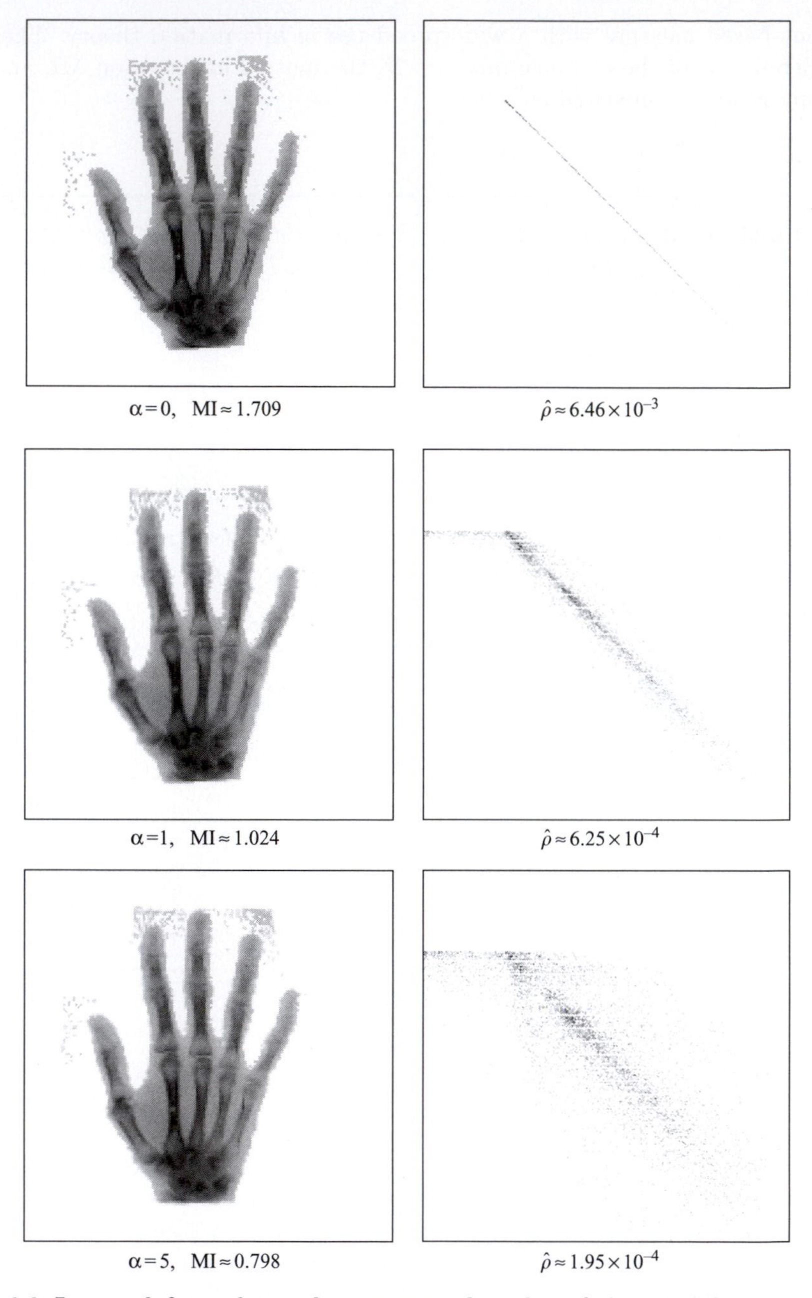

FIG. 6.2 LEFT: deformed template, RIGHT: log-plot of the joint density ρ_{T,T_φ}, where φ is a rotation of $\alpha, \alpha = 0, 1$, and 5 degrees, $\text{MI} := \mathcal{D}^{\text{MI}}[T, T \circ \varphi]$, $\hat{\rho} := \|\log(1 + \hat{\rho}_{T,T_\alpha})\|_\infty$. For illustration purposes, the density has been re-scaled and background values have been neglected.

since

$$
\begin{aligned}
&\mathbb{E}_{\rho_{R,T}}\left[\log \frac{\rho_{R,T}}{\rho_R \rho_T}\right] \\
&= \int_{\mathbb{R}^2} (\log \rho_{R,T}(g_1, g_2) - \log \rho_R(g_1) - \log \rho_T(g_2))\rho_{R,T}(g_1, g_2)\, d(g_1, g_2) \\
&= -H(\rho_{R,T}) - \int_{\mathbb{R}} \int_{\mathbb{R}} \rho_{R,T}(g_1, g_2)\, dg_2 \, \log \rho_R(g_1)\, dg_1 \\
&\quad - \int_{\mathbb{R}} \int_{\mathbb{R}} \rho_{R,T}(g_1, g_2)\, dg_1 \, \log \rho_T(g_2)\, dg_2 \\
&= -H(\rho_{R,T}) - \int_{\mathbb{R}} \rho_R(g_1) \log \rho_R(g_1)\, dg_1 - \int_{\mathbb{R}} \rho_T(g_2) \log \rho_T(g_2)\, dg_2,
\end{aligned}
$$

where Fubini's theorem has been used.

However, we are interested in a differentiable distance measure, and for non-smooth gray value densities, the mutual information is not differentiable. In order to circumvent this disadvantage, we approximate the joint density by a smooth, i.e., differentiable, approximation using Parzen window techniques.

Here we summarize the approach presented in Wells et al (1996). The basic idea is to estimate the density ρ_S using a given sample $X = (X_1, \ldots, X_n)$ and the entropy H_S given a sample $Y = (Y_1, \ldots, Y_m)$, where $X_j \in \Omega$ and $Y_k \in \Omega$ are independently and identically distributed (i.i.d.) on Ω. The mutual information, which is not directly accessible, is computed in terms of these estimates.

In the first step, the densities ρ_R, ρ_T, and $\rho_{R,T}$ are estimated using the so-called *Parzen window method* (see, e.g., Duda & Hart (1973)),

$$\rho_S(z) \approx \rho_S(z, X) := \frac{1}{n} \sum_{j=1}^{n} g_q\big(\Sigma_S, (z - S(X_j))\big), \tag{6.11}$$

where $g_q : \mathbb{R}^{q \times q} \times \mathbb{R}^q \to \mathbb{R}$ is chosen to be a q-variate Gaussian density

$$g_q(\Sigma, z) := (2\pi)^{-q/2} (\det \Sigma)^{-1/2} \exp(-z^\top \Sigma^{-1} z/2), \tag{6.12}$$

i.e., $\Sigma \in \mathbb{R}^{d \times d}$ and $z \in \mathbb{R}^d$. For $S = R, T$, we have $q = 1$, and for $S = (R, T)^\top$, we have $q = 2$. The question of how to find the optimal values for Σ is discussed later. Note that Σ_R, Σ_T, and $\Sigma_{(R,T)^\top}$ are the covariance of the Parzen window and not of the density to be estimated.

In the second step, the entropies are estimated using a Monte Carlo method, i.e.,

$$H(\rho_S) = \mathbb{E}_{\rho_S}\left[-\log \rho_S\right] \approx -\frac{1}{m}\sum_{k=1}^{m} \log \rho_S(S(Y_k)).$$

Combining these two approximation steps, one obtains the following estimates $\tilde{H}(\rho_S; X, Y)$ for the entropies $H(\rho_S)$,

$$\begin{aligned} H(\rho_S) &\approx -\frac{1}{m}\sum_{k=1}^{m} \log \rho_S(S(Y_k)) \\ &\approx -\frac{1}{m}\sum_{k=1}^{m} \log \rho_S(S(Y_k), X) \\ &= -\frac{1}{m}\sum_{k=1}^{m} \log\left(\frac{1}{n}\textstyle\sum_{j=1}^{n} g_q\big(\Sigma_S, \big(S(Y_k) - S(X_j)\big)\big)\right) \\ &=: \tilde{H}(S; X, Y). \end{aligned}$$

In Viola (1995, p. 49f.) it is argued that a measure for the quality of the Parzen estimate is to evaluate the standard deviation normalized by the mean. Viola gives the estimation

$$\frac{\sigma(P^*(x, X))}{\mathbb{E}\left[P^*(x, X)\right]} \approx \sqrt{\frac{k}{n}}\sqrt{\frac{k - P^*(x, X)}{P^*(x, X)}},$$

where P^* is the Parzen window approximation based on the sample X, n is the length of the sample, and k is a normalization constant, chosen such that P^* integrates to one; cf., Viola (1995, p. 50). Note that this estimate neglects the dependence of the density on the bandwidth σ. Considering also this dependency, the rate of convergence is only $n^{-2/5}$, following standard arguments; cf., Dümbgen (2001).

Estimating a univariate density ρ from the i.i.d. sample $X_j, j = 1, \ldots, n$, using the kernel estimator

$$\hat{\rho}_{n,\sigma}(y) := \frac{1}{n}\sum_{j=1}^{n} g(\sigma, y - X_j),$$

where $g = g_1$, we have

$$\begin{aligned}
\mathbb{E}_\rho\left[\hat{\rho}_{n,\sigma}(x)\right] &= \frac{1}{n}\mathbb{E}_\rho\left[\sum_{j=1}^{n} g(\sigma, x - X_j)\right] = \mathbb{E}_\rho\left[g(\sigma, x - X_1)\right] \\
&= \int_{\mathbb{R}} g(\sigma, x - y)\rho(y)dy \\
&= \int_{\mathbb{R}} \frac{1}{\sqrt{2\pi\sigma}} e^{-(x-y)^2/(2\sigma)}\rho(y)dy \\
&= \frac{1}{\sqrt{2\pi}} \int_{\mathbb{R}} e^{-z^2/2}\rho(x - \sqrt{\sigma}z)dz \\
&= \frac{1}{\sqrt{2\pi}} \int_{\mathbb{R}} e^{-z^2/2}\big(\rho(x) - \sqrt{\sigma}\rho'(x)z + \sigma\rho''(x - \theta z)z^2\big)dz \\
&= \rho(x) + \mathcal{O}(\sigma), \\
\mathrm{Var}_\rho[\hat{\rho}_{n,\sigma}(x)] &= \frac{1}{n}\mathrm{Var}_\rho[g(\sigma, x - X_1)] \\
&= \frac{1}{n}\int_{\mathbb{R}} g(\sigma, x - y)^2\rho(y)dy \\
&= \frac{1}{n}\int_{\mathbb{R}} \frac{1}{2\pi\sigma} e^{-(x-y)^2/\sigma}\rho(y)dy \\
&= \frac{1}{n\sqrt{\sigma}}\int_{\mathbb{R}} \frac{1}{2\pi} e^{-z^2}\rho(x - \sqrt{\sigma}z)dz \\
&= \mathcal{O}((n\sqrt{\sigma})^{-1}).
\end{aligned}$$

Hence,

$$\begin{aligned}
&\mathbb{E}_\rho\left[(\hat{\rho}_{n,\sigma}(x) - \rho(x))^2\right]^{-1/2} \\
&\quad = \big((\mathbb{E}_\rho\left[\hat{\rho}_{n,\sigma}(x)\right] - \rho(x))^2 + \mathrm{Var}_\rho[\hat{\rho}_{n,\sigma}(x)]\big)^{-1/2} \\
&\quad = \big(\mathcal{O}(\sigma^2) + \mathcal{O}((n\sqrt{\sigma})^{-1})\big)^{-1/2} \\
&\quad = \mathcal{O}(n^{-2/5}), \quad \text{for optimal } \sigma = C \cdot n^{-2/5}.
\end{aligned}$$

In Viola (1995), no analysis with respect to the approximation order of the Monte Carlo approach for the entropy estimation is given.

For the choices of the variances in the Parzen window functions, Viola suggests a cross-validation approach based on the sample X. However, concrete formulas and the approximation order of these estimations are missing. Moreover, the assumption that the covariance matrix Σ_{R,T_a} of the Parzen window for the joint density $S = (R, T_a)^\top$ is diagonal is a severe restriction on the images under consideration.

It is worthwhile noticing that modifications of the mutual information may also be used. Studholme et al (1996) for example use

$$\mathcal{D}^{\mathrm{MIN}}[R,T] := \frac{H(\rho_{R,T})}{H(\rho_T)+H(\rho_R)}.$$

6.4.2 *Gradient of mutual information*

For the computation of the Gâteaux derivative of the Parzen window estimation-based approximation of the mutual information $\mathcal{D}^{\mathrm{MI}}[R,T;\varphi]$ with respect to a perturbation ψ, we introduce the following abbreviations:

$$\begin{aligned}
T[\varphi] := T\circ\varphi, \quad \sigma := \Sigma_{T[\varphi]}\in\mathbb{R}, \quad H_T[\varphi] := H(T[\varphi]),\\
g_{j,k}[\varphi] := g_1\big(\sigma,\big(T[\varphi](Y_k)-T[\varphi](X_j)\big)\big),\\
\Sigma := \Sigma_{(R,T[\varphi])^\top}\in\mathbb{R}^{2\times 2}, \quad H_{R,T}[\varphi] := H((R,T[\varphi])^\top),\\
G_{j,k}[\varphi] := g_2\big(\Sigma,\big(R(Y_k)-R(X_j),T[\varphi](Y_k)-T[\varphi](X_j)\big)^\top\big),
\end{aligned}$$

where σ and Σ are assumed to be independent on φ. Hence

$$\begin{aligned}
dT[\varphi;\psi] &= \left\langle \nabla T\big|_{\varphi},\psi\right\rangle_{\mathbb{R}^d},\\
g_{j,k}[\varphi] &= \frac{1}{\sqrt{2\pi\sigma}}\cdot\exp\big(-(T[\varphi](Y_k)-T[\varphi](X_j))^2/(2\sigma)\big),\\
dg_{j,k}[\varphi;\psi] &= -g_{j,k}[\varphi]\big(T[\varphi](Y_k)-T[\varphi](X_j)\big)\cdot\sigma^{-1}\big(dT[\varphi;\psi](Y_k)-dT[\varphi;\psi](X_j)\big),\\
H_T[\varphi] &= \frac{1}{m}\sum_{k=1}^{m}\log\left(\frac{1}{n}\sum_{j=1}^{n}g_{j,k}[\varphi]\right),\\
dH_T[\varphi;\psi] &= \frac{1}{m}\sum_{k=1}^{m}\sum_{j=1}^{n}\frac{dg_{j,k}[\varphi;\psi]}{\sum_{\ell=1}^{n}g_{\ell,k}[\varphi]},\\
G_{j,k}[\varphi] &= \frac{1}{2\pi\sqrt{\det(\Sigma)}}\\
&\quad\cdot\exp\left(-\frac{1}{2}\begin{pmatrix}R(Y_k)-R(X_j)\\ T[\varphi](Y_k)-T[\varphi](X_j)\end{pmatrix}^\top\cdot\Sigma^{-1}\begin{pmatrix}R(Y_k)-R(X_j)\\ T[\varphi](Y_k)-T[\varphi](X_j)\end{pmatrix}\right),
\end{aligned}$$

$$dG_{j,k}[\varphi;\psi] = -G_{j,k}[\varphi] \begin{pmatrix} R(Y_k) - R(X_j) \\ T[\varphi](Y_k) - T[\varphi](X_j) \end{pmatrix}^\top$$

$$\cdot \Sigma^{-1} \begin{pmatrix} 0 \\ dT[\varphi;\psi](Y_k) - dT[\varphi;\psi](X_j) \end{pmatrix},$$

$$H_{R,T}[\varphi;\psi] = \frac{1}{m} \sum_{k=1}^{m} \log \left(\frac{1}{n} \sum_{j=1}^{n} G_{j,k}[\varphi] \right),$$

$$dH_{R,T}[\varphi;\psi] = \frac{1}{m} \sum_{k=1}^{m} \sum_{j=1}^{n} \frac{dG_{j,k}[\varphi;\psi]}{\sum_{\ell=1}^{n} G_{\ell,k}[\varphi]}.$$

Assuming that $\Sigma = \Sigma_{(R,T[\varphi])^\top}$ is also diagonal, $\Sigma = \operatorname{diag}(\sigma_1, \sigma_2)$, we finally obtain

$$\begin{aligned} d\mathcal{D}^{\mathrm{MI}}[R,T;\varphi;\psi] &= dH_T[\varphi;\psi] - dH_{R,T}[\varphi;\psi] \\ &= \frac{1}{m} \sum_{k=1}^{m} \sum_{j=1}^{n} \left(\frac{dg_{j,k}[\varphi;\psi]}{\sum_{\ell=1}^{n} g_{\ell,k}[\varphi]} - \frac{dG_{j,k}[\varphi;\psi]}{\sum_{\ell=1}^{n} G_{\ell,k}[\varphi]} \right) \\ &= \frac{1}{m} \sum_{k=1}^{m} \sum_{j=1}^{n} \left(\frac{g_{j,k}[\varphi]}{\sigma \sum_{\ell=1}^{n} g_{\ell,k}[\varphi]} - \frac{G_{j,k}[\varphi]}{\sigma_2 \sum_{\ell=1}^{n} G_{\ell,k}[\varphi]} \right) \\ &\quad \cdot \big(T[\varphi](Y_k) - T[\varphi](X_j)\big)\big(dT[\varphi;\psi](Y_k) - dT[\varphi;\psi](X_j)\big). \end{aligned}$$

If in particular the transformation φ is parametric, i.e., $\varphi = \varphi_a$, the derivatives can be computed explicitly.

Example 6.1 For an affine linear registration based on mutual information we set

$$\varphi_a(x) = \begin{pmatrix} a_1 & a_2 \\ a_3 & a_4 \end{pmatrix} \begin{pmatrix} x_1 \\ x_2 \end{pmatrix} + \begin{pmatrix} a_5 \\ a_6 \end{pmatrix}$$

and with

$$\psi_1(x) = (b_1 x_1, 0)^\top, \quad \psi_2(x) = (b_2 x_2, 0)^\top, \quad \psi_3(x) = (0, b_3 x_1)^\top,$$

$$\psi_4(x) = (0, b_4 x_2)^\top, \quad \psi_5(x) = (b_5, 0)^\top, \quad \psi_6(x) = (0, b_6)^\top,$$

$$\Delta w_{j,k}(a) := \frac{g_{j,k}[\varphi_a]}{\sigma \sum_{\ell=1}^{n} g_{\ell,k}[\varphi_a]} - \frac{G_{j,k}[\varphi_a]}{\sigma_2 \sum_{\ell=1}^{n} G_{\ell,k}[\varphi_a]},$$

$$\Delta T_{j,k}(a) := T(\varphi_a(Y_k)) - T(\varphi_a(X_j)),$$

we thus have

$$\partial_{a_1} D^{\mathrm{MI}}(R,T;a) = d\mathcal{D}^{\mathrm{MI}}[R,T;\varphi_a,\psi_1]$$
$$= \frac{1}{m}\sum_{k=1}^{m}\sum_{j=1}^{n} \Delta w_{j,k}(a)\Delta T_{j,k}(a)\big(\partial_{x_1}T(\varphi_a(Y_k))Y_{k,1} - \partial_{x_1}T(\varphi(X_j))X_{j,1}\big),$$

$$\partial_{a_2} D^{\mathrm{MI}}(R,T;a) = d\mathcal{D}^{\mathrm{MI}}[R,T;\varphi_a,\psi_2]$$
$$= \frac{1}{m}\sum_{k=1}^{m}\sum_{j=1}^{n} \Delta w_{j,k}(a)\Delta T_{j,k}(a)\big(\partial_{x_1}T(\varphi_a(Y_k))Y_{k,2} - \partial_{x_1}T(\varphi(X_j))X_{j,2}\big),$$

$$\partial_{a_3} D^{\mathrm{MI}}(R,T;a) = d\mathcal{D}^{\mathrm{MI}}[R,T;\varphi_a,\psi_3]$$
$$= \frac{1}{m}\sum_{k=1}^{m}\sum_{j=1}^{n} \Delta w_{j,k}(a)\Delta T_{j,k}(a)\big(\partial_{x_2}T(\varphi_a(Y_k))Y_{k,1} - \partial_{x_2}T(\varphi(X_j))X_{j,1}\big),$$

$$\partial_{a_4} D^{\mathrm{MI}}(R,T;a) = d\mathcal{D}^{\mathrm{MI}}[R,T;\varphi_a,\psi_4]$$
$$= \frac{1}{m}\sum_{k=1}^{m}\sum_{j=1}^{n} \Delta w_{j,k}(a)\Delta T_{j,k}(a)\big(\partial_{x_2}T(\varphi_a(Y_k))Y_{k,2} - \partial_{x_2}T(\varphi(X_j))X_{j,2}\big),$$

$$\partial_{a_5} D^{\mathrm{MI}}(R,T;a) = d\mathcal{D}^{\mathrm{MI}}[R,T;\varphi_a,\psi_5]$$
$$= \frac{1}{m}\sum_{k=1}^{m}\sum_{j=1}^{n} \Delta w_{j,k}(a)\Delta T_{j,k}(a)\big(\partial_{x_1}T(\varphi_a(Y_k)) - \partial_{x_1}T(\varphi(X_j))\big),$$

$$\partial_{a_6} D^{\mathrm{MI}}(R,T;a) = d\mathcal{D}^{\mathrm{MI}}[R,T;\varphi_a,\psi_6]$$
$$= \frac{1}{m}\sum_{k=1}^{m}\sum_{j=1}^{n} \Delta w_{j,k}(a)\Delta T_{j,k}(a)\big(\partial_{x_2}T(\varphi_a(Y_k)) - \partial_{x_2}T(\varphi(X_j))\big),$$

where $X_j = (X_{j,1}, X_{j,2})^\top$ and $Y_k = (Y_{k,1}, Y_{k,2})^\top$, respectively.

Wells et al (1996) proposed a gradient-based steepest descent method with fixed step size α for the minimization of D; cf., Algorithm 6.8.

Algorithm 6.8 *Stochastic maximization algorithm of Wells et al (1996).*
Repeat:
Collect sample $X = (X_1, \ldots, X_{n_x})$ from Ω.
Collect sample $Y = (Y_1, \ldots, Y_{n_y})$ from Ω.
Update $a \mapsto a + \lambda \nabla_a D(a)$.

The steepest descent method in Algorithm 6.8 is based on different samples in different iteration steps. This implies that the objective function changes from step to step. A convergence proof for the modified algorithm is missing. Finally, as is well-known, the convergence rate of the steepest descent method might be arbitrarily slow, a fact that has already been observed for quadratic optimization problems; cf., e.g., Fletcher (1987).

Our implementation is based on a Levenberg–Marquardt-type technique; cf., Algorithm 6.9. To this end, let $f : \mathbb{R}^n \to \mathbb{R}$, $f(a) := -D^{\mathrm{MI}}(R, T; \varphi_a)$. For a given a and λ, we replace f by the quadratic model

$$q_a(\Delta a) := f(a) + \nabla f(a)\Delta a + \tfrac{1}{2}\Delta a[\nabla f(a)\nabla f(a)^\top]\Delta a,$$

where $[\nabla f(a)\nabla f(a)^\top]$ gives an approximation to the Hessian matrix $\nabla^2 f(a)$.

Remark 6.1 Let $f : \mathbb{R}^n \to \mathbb{R}^m$ be a function and $h : \mathbb{R}^n \to \mathbb{R}$ be defined by $h(a) = \frac{1}{2}\|f(a)\|^2_{\mathbb{R}^m}$. The $(p, q)^{th}$ entry of the Hessian matrix of h is given by

$$\frac{\partial^2 h}{\partial a_p \partial a_q}(a) = \sum_{j=1}^{m} \left(\frac{\partial f_j}{\partial a_p}(a) \frac{\partial f_j}{\partial a_q}(a) + f_j(a) \frac{\partial^2 f_j}{\partial a_p \partial a_q}(a) \right).$$

If $\|f(a)\|_{\mathbb{R}^m}$ is small enough, we have $\nabla^2 h(a) \approx \nabla f(a)\nabla f(a)^\top$; see also Nocedal & Wright (1999, §10.2).

The next step is the minimization of the quadratic model, subject to $\|\Delta a\|_{\mathbb{R}^n} \le h$. The restriction results in

$$[\nabla f(a)\nabla f(a)^\top + \lambda I_n]\Delta a = -\nabla f(a),$$

where h and λ are related, and for a meaningful choice of h we have that $[\nabla f(a)\nabla f(a)^\top + \lambda I_n]$ is positive definite; see, e.g., Fletcher (1987, §5.2). The next iterate is given by $a' := a + \Delta a$. We denote the iterates in the kth iteration by $a^{(k)}$ and $\lambda^{(k)}$, respectively. If $f(a^{(k+1)} \ge f(a^{(k)})$, the kth step is not successful. In this situation, we increase the value of $\lambda^{(k)}$, i.e., shrink the size h of the trust region in our model; see Algorithm 6.9 for details.

Note that the objective function f also depends on the samples X and Y which change from step to step. Thus, we measure convergence numerically by the variance of the last m parameters $a^{(k-m+1)}, \ldots, a^{(k)}$, where k denotes the iteration and typically $m = 10$.

6.5 An example of mutual information-based affine linear registration

The result of a mutual information-based optimal affine linear registration is shown in Fig. 6.3. Here, the intensity values of the template image have been modified. The gray values of the initial image have been re-scaled and inverted.

Algorithm 6.9 *Minimization of Parzen window approximation of mutual information $f(a) := -D^{\mathrm{MI}}(R, T; \varphi_a)$ with respect to the parameter a using a Levenberg–Marquardt-type technique.*

> Set $\mu = 1$, set $\beta > 1$, e.g., $\beta = 5$. Choose initial $a \in \mathbb{R}^n$, e.g., $a = (1, 0, 0, 1, 0, 0)^\top$ for affine linear registration.
> Repeat
>   Compute $f(a)$ and $\nabla f(a)$, set $\lambda = \mu$.
>   Compute Δ_a from $[\nabla f(a) \nabla f(a)^\top + \lambda I_n] \Delta_a = -\nabla f(a)$.
>   While $f(a + \Delta a) \geq f(a)$,
> $\lambda = \beta\lambda$.
>   Compute Δ_a from $[\nabla f(a) \nabla f(a)^\top + \lambda I_n] \Delta_a = -\nabla f(a)$,
>   end.
>   Set $\mu = \lambda/\beta$, $a = a + \Delta_a$.
> Until convergence.

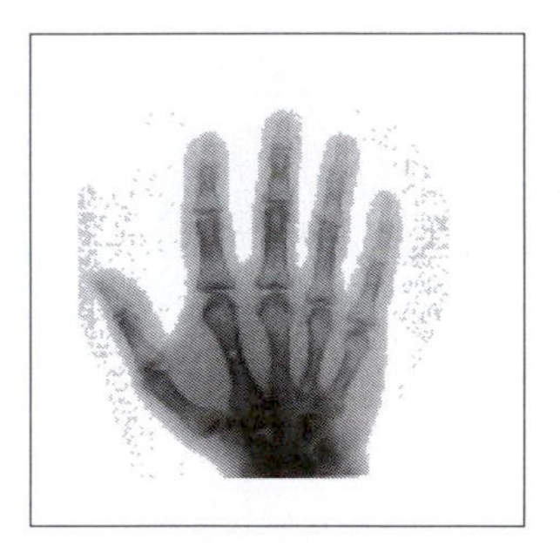
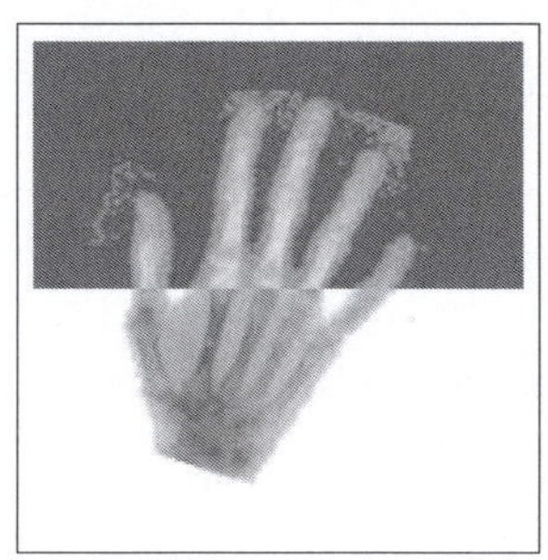
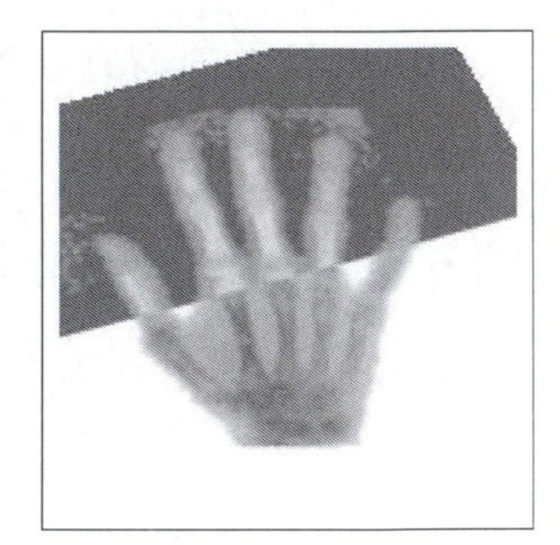

FIG. 6.3 Optimal mutual information-based affine linear registration; reference (LEFT), template (MIDDLE), template after registration (RIGHT).

FIG. 6.4 Intensity and mutual information-based registration.

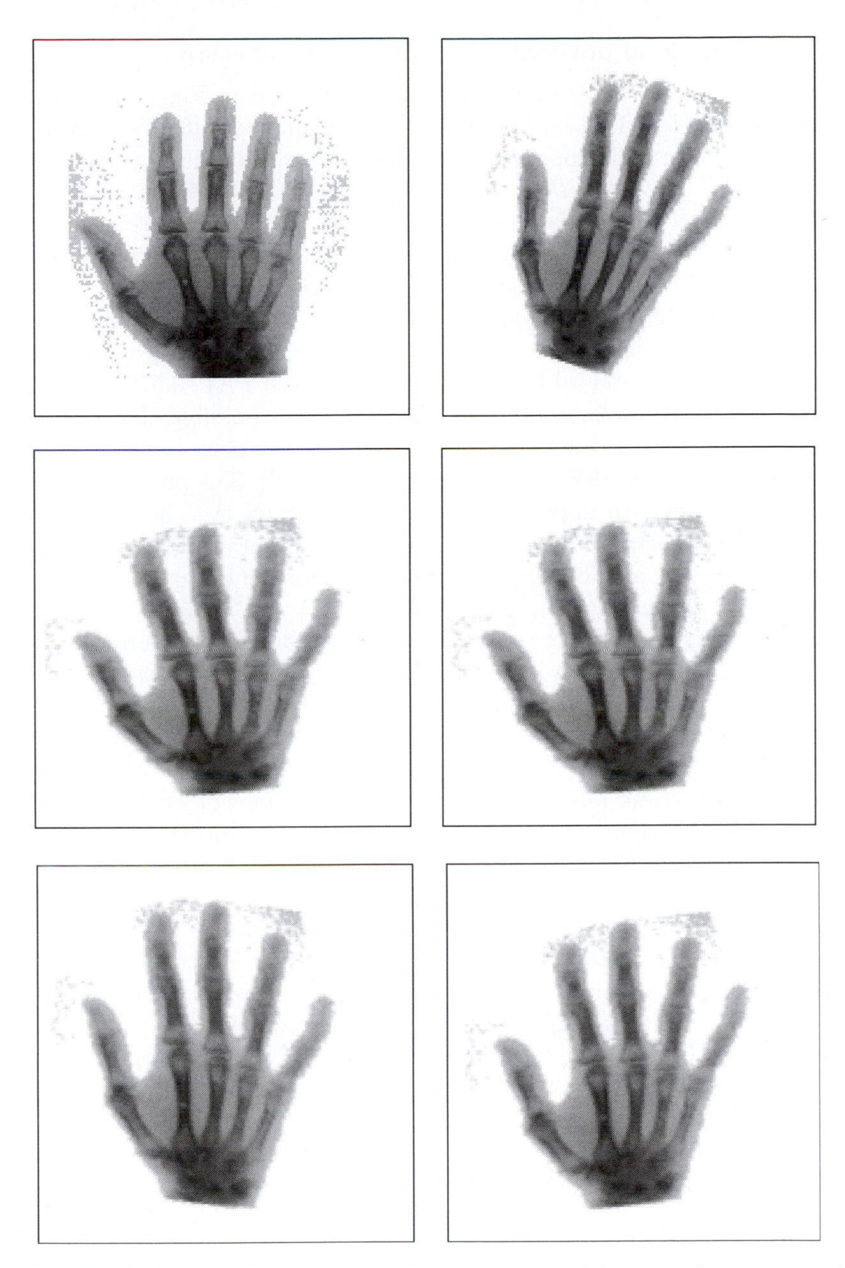

FIG. 6.5 Four different linear registrations: reference (TOP LEFT), template (TOP RIGHT), standard PAT registration, (MIDDLE LEFT), robust PAT registration (MIDDLE RIGHT), affine linear SSD registration (BOTTOM LEFT), and affine linear MI registration (BOTTOM RIGHT).

6.6 An example of optimal affine linear registration

Four different linear registration results are presented in Fig. 6.5. In order to compare the standard PAT, robust PAT, sum of squared differences (SSD), and mutual information (MI) approaches, we use images sharing the same modality. Measuring $D^{\mathrm{SSD}}(R,T;a^*)/\mathcal{D}^{\mathrm{SSD}}[R,T]$, where a^* is the optimal parameter set obtained from the different approaches, we have the following results: 56% for standard PAT, 57% for robust PAT, 51% for optimized SSD, and 62% for optimized MI. Note that our MI computation is based on an approximation based on random variables. Thus a comparison with respect to MI has a stochastic component. For one particular measurement of $D^{\mathrm{MI}}(R,T;a^*)$ we get the following results: 0.4812% for standard PAT, 0.4682% for robust PAT, 0.5106% for optimized SSD, and 0.5035% for optimized MI. Thus, it is possible that the optimal MI solution is suboptimal with respect to different samples.

A comparison of the different techniques is difficult. The intensity-based distance measure seems to appear natural to the human eye. If the gray values of the images are related, this measure gives visually pleasing results. However, if there is no simple relation between gray values of the images, intensity-based registration is certainly not the best choice.

The irritating point about mutual information is that it does not necessarily match intensities. Figure 6.4 illustrates this phenomenon. Suppose we want to register the two images displayed in this figure. Using intensity-based linear registration, we find two minima, i.e., rotation of $k\pi/2, k = 1,3$. For this solutions, the intensities of the reference and mapped template images coincide. Using mutual information, we find four solutions, i.e., rotations of $k\pi/2$, $k = 0,1,2,3$.

Minimization becomes a delicate matter, since the objective function is not convex, as illustrated by the above example. Avoiding convergence to local minima requires additional techniques, such as a multi-scale approach. For intensity-based registration, typically a Gauss pyramid is used. For mutual information, one may also view the sample size as a scale-space parameter. A direct comparison is impossible, since fast numerical schemes for mutual information are always based on this additional scale-space parameter.

7

SUMMARY OF PARAMETRIC IMAGE REGISTRATION

Different parametric image registration techniques have been discussed. The techniques are all based on the minimization of a certain distance measure, and the distance measure is based on image features or directly on image intensities. Image features can be user supplied (e.g., so-called landmarks) or may be deduced automatically from the image intensities (e.g., so-called principal axes). Typical examples of intensity-based distance measures are the sum of squared differences (cf., Definition 6.1), correlation (cf., Definition 6.5), or mutual information (cf., Definition 6.7).

For all proposed techniques, the transformation is parametric, i.e., it can be expanded in terms of some parameters α_j and basis functions ψ_j. The required transformation is a minimizer of the distance measure in the space spanned by the basis functions ψ_j, $j = 1, \dots, n$. The minimizer can be obtained from some algebraic equations or by applying appropriate optimization tools.

Landmark-based parametric registration

- Supply features $\mathcal{F}(R, j) = x^{R,j}$ and $\mathcal{F}(T, j) = x^{T,j}$, $j = 1, \dots, m$. Choose a set of basis functions. Find parameters $\alpha = (\alpha_1, \dots, \alpha_n)^\top \in \mathbb{R}^n$ such that for $\varphi = \sum_{j=1}^{n} \alpha_j \psi_j$,

$$\mathcal{D}^{\mathrm{LM}}[\varphi] = \sum_{j=1}^{m} \|\mathcal{F}(R, j) - \varphi(\mathcal{F}(T, j))\|_f = \min.$$

  The solution is given by algebraic equations for the coefficients; cf., Section 4.2.
- Needs landmarks.
- Simple.
- Only needs the numerical solution of a linear system of equations.
- Least squares matrix may not have full rank; implicit and in general unknown additional conditions on the features.
- Results might be arbitrarily awful.

Landmark-based smooth registration

- Supply features $\mathcal{F}(R, j) = x^{R,j}$ and $\mathcal{F}(T, j) = x^{T,j}$, $j = 1, \dots, m$. Choose a regularizer $\mathcal{S}^{\mathrm{TPS}}$ (cf., eqn (4.22)) and a regularizing parameter $\alpha \geq 0$.

For $\alpha = 0$, find φ such that

$$\mathcal{S}^{\mathrm{TPS}}[\varphi] = \min \text{ subject to } \varphi(\mathcal{F}(T,j)) = \mathcal{F}(R,j), \quad j = 1, \ldots, m.$$

Alternatively, for $\alpha > 0$, find φ such that

$$\alpha \mathcal{S}^{\mathrm{TPS}}[\varphi] + \mathcal{D}^{\mathrm{LM}}[\varphi] = \min .$$

The solutions are given by algebraic equations for the coefficients in a radial basis expansion; cf., Section 4.3.4.

- Needs landmarks.
- Only needs the numerical solution of a linear system of equations, essentially m unknowns and m equations; system is always non-singular.
- Physically meaningful transformation, minimizes curvature.
- Results may be bad.

Principal axes-based registration

- Compute $\varphi \in \Pi_1^d(\mathbb{R}^d)$, such that $\mathcal{F}(T \circ \varphi) = \mathcal{F}_R$, where $\mathcal{F}_B$ is a feature vector containing the center of gravity, the standard deviations, and the principal axis based on an appropriate density class, e.g. Gauss or Cauchy density; cf. Chapter 5. A solution can be deduced from an eigenvalue decomposition of the moment matrices of R and T, respectively; cf., Theorem 5.2.
- Simple, fast, easy to understand and to interpret.
- Needs moment matrix and eigenvalue decompositions of two d-by-d matrices.
- Not suitable for multimodal densities/images.
- Ambiguous results.
- Very few registration parameters.

Optimal parametric registration

- Choose an appropriate distance measure $\mathcal{D}$. Choose a set of basis functions. Find parameters $\alpha = (\alpha_1, \ldots, \alpha_n)^\top \in \mathbb{R}^n$ such that for $\varphi = \sum_{j=1}^n \alpha_j \psi_j$,

$$\mathcal{D}[\varphi] = \min .$$

A numerical solution can be obtained by using optimization methods, e.g., a Gauss–Newton or Levenberg–Marquardt method.

- General, flexible.
- No physical, meaningful transformation.
- Optimization can be very slow, in particular for high-dimensional spline spaces.

Part II

Non-parametric image registration

8

NON-PARAMETRIC IMAGE REGISTRATION

The approaches discussed in the previous part are based on certain parameters. Either the transformation φ is parametric, i.e., can be expanded in terms of some basis functions, or the registration is driven by external features, e.g., the so-called *landmarks*; cf., Chapter 4. The choice of the basis functions is in general artificial and needs to be discussed and justified. The detection of landmarks is still a delicate matter and requires additional time and expert knowledge; see, e.g. Rohr (2001).

In this part, non-parametric registration techniques are to be discussed. The idea behind this type of registration is to come up with an appropriate measure both for the similarity of images as well as for the likelihood of a non-parametric transformation.

The purpose of this chapter is to set up a general framework for consideration of different registration techniques. This framework is based on a variational formulation of the registration problem, and the numerical schemes to be considered are based on the Euler–Lagrange equations which characterize a minimizer. This general concept will be specified for various registration techniques in the next chapters.

8.1 A general framework

Given two images, a reference R and a template T, we are looking for a transformation φ, such that T_φ, where $T_\varphi(x) := T(\varphi(x))$, is similar to R in a certain sense. For the following discussion it is also convenient to split the transformation φ into the trivial identity part and the so-called *deformation* or *displacement* u, $u : \mathbb{R}^d \to \mathbb{R}^d$, i.e.,

$$\varphi(x) = x - u(x), \tag{8.1}$$

where we exploit an Eulerian viewpoint; cf., Section 3.3.2. To be precise, let $\hat{x}$, $\hat{u}$, and $\hat{\varphi}$ denote the point, displacement, and transformation with respect to the Lagrange coordinates and let x, u, and φ denote the point, displacement, and transformation with respect to the Euler coordinates. Thus,

$$x = \hat{\varphi}(\hat{x}) = \hat{x} + \hat{u}(\hat{x}) \quad \text{and} \quad \hat{x} = \varphi(x) = x - u(x),$$

where $\varphi = \hat{\varphi}^{-1}$, $\hat{u}(\hat{x}) = u(x)$, and φ is assumed to be diffeomorphic.

The most intuitive way of approaching the registration problem is to design a suitable distance measure $\mathcal{D}$ and to minimize the distance between R and T_u with respect to u,

$$\mathcal{D}[R,T;u] := \mathcal{D}[R,T_u] \xrightarrow{u} \min, \tag{8.2}$$

where, in order to keep the notation simple, we introduce

$$T_u(x) := T(x - u(x)). \tag{8.3}$$

A direct minimization of the distance measure has some drawbacks: the problem is ill-posed since small changes of the input data may lead to large changes of the output data, the solution is not unique since the problem is not convex, and the deformation may not even be continuous. Thus, it is not possible to construct an appropriate scheme for a numerical solution.

Moreover, in many applications additional implicit assumptions are made on the transformation. For example, within the HNSP it is implicitly assumed that the optimal transformation is diffeomorphic. This implies that no additional cracks and/or folding of the tissue are introduced by the registration process.

The remedy in both situations is to add an additional *regularizing* term or *smoother* $\mathcal{S}$. With an appropriate smoother it becomes possible to distinguish particular transformations which seem to be more likely than others. Moreover, the regularized problem becomes a starting point for a stable numerical implementation. The choice of the smoother depends on the particular application. In many applications the desired properties of the transformation are not known a priori. Therefore, different smoothing techniques have to be provided.

Problem 8.1 *Given two images R, T, and a positive regularizing parameter $\alpha \in \mathbb{R}_{>0}$, find a deformation u, such that*

$$\mathcal{J}[u] := \mathcal{D}[R,T;u] + \alpha\mathcal{S}[u] = \min .$$

Typical choices of smoothers $\mathcal{S}$ to be discussed in the following chapters are based on bi-linear forms a, i.e.,

$$\mathcal{S}[u] = \tfrac{1}{2}a[u,u],$$

where the bi-linear forms can be traced back to the inner product in $L_2(\mathbb{R}^d)$.

A necessary condition for a minimizer u of Problem 8.1 is that the Gâteaux derivative $d\mathcal{J}[u;v]$ of $\mathcal{J}$ vanishes for all suitable perturbations v. This derivative was also known as the first variation of $\mathcal{J}$ in the direction of v.

For the Gâteaux derivative of $\mathcal{S}$, we will find

$$\begin{aligned} d\mathcal{S}[u;v] &= \lim_{h\to 0}\frac{1}{2h}\big(a[u+hv,u+hv] - a[u,u]\big) = a[u,v] \\ &= \int_{\mathbb{R}^d} \langle \mathcal{A}[u](x), v(x)\rangle_{\mathbb{R}^d}\, dx, \end{aligned} \tag{8.4}$$

where $\mathcal{A}$ is a partial differential operator. For the last equality, we apply Green's formula; cf., e.g., Buck (1978, §9.4). When appropriate, additional boundary conditions have to be imposed such that boundary integrals vanish.

Finally, for the Gâteaux derivative of $\mathcal{D}$, we will find

$$d\mathcal{D}[u;v] = \int_{\mathbb{R}^d} \langle f(x, u(x)), v(x) \rangle_{\mathbb{R}^d} \, dx,$$

where the so-called *force* f depends on the particular distance measure.

The minimization problem 8.1 has been formulated without explicit boundary conditions on a minimizer u, such that implicit boundary conditions arise naturally. However, due to our experiments, boundary conditions are of minor importance if the images are pre-registered and embedded into a uniform background. Thus, we may use explicit boundary conditions to come up with efficient numerical schemes.

Before starting the discussion of particular choices for distance measures $\mathcal{D}$ and smoothers $\mathcal{S}$ we make some general comments.

8.2 General solution schemes

A variety of different numerical schemes for the computation of a numerical solution of Problem 8.1 can be found in the literature. A first choice is a gradient-based steepest descent method. Henn (1997) used a scheme based on the gradient of the functional in Problem 8.1, while the scheme of Thirion (1995) might also be viewed as a particular gradient scheme. Here, the gradient of $\mathcal{D}$ is computed and projected in a suitable space. Higher order schemes, e.g., Newton-type schemes, cannot be recommended, since these schemes are in need of higher order derivatives of the image T, and the computation of higher order derivatives is in general time consuming and numerically unstable.

The class of approaches to be discussed here is based on the characterization of a minimizer by the Euler–Lagrange equations for Problem 8.1, which are of the form

$$\mathcal{A}[u](x) - f(x, u(x)) = 0, \quad \text{for all } x \in \Omega, \tag{8.5}$$

where $\Omega =]0,1[^d$ is the region under consideration and d denotes the spatial dimension of the images. The partial differential operator $\mathcal{A}$ is related to the smoother $\mathcal{S}$ (see also eqn (8.4)), and the *force* f is related to the distance measure $\mathcal{D}$.

A convenient way of solving this semi-linear partial differential equation and to by-pass the non-linearity is to exploit a fixed-point iteration scheme. Starting with an initial guess $u^{(0)}$ (e.g., $u^{(0)} \equiv 0$), we define $u^{(k+1)}$ implicitly by

$$\mathcal{A}[u^{(k+1)}](x) = f(x, u^{(k)}(x)), \quad x \in \Omega, \quad k \in \mathbb{N}_0. \tag{8.6}$$

The following slight modification of this iteration can be used to stabilize the scheme,

$$u^{(k+1)}(x) + \tau\mathcal{A}[u^{(k+1)}](x) = u^{(k)}(x) + \tau f(x, u^{(k)}(x)), \quad x \in \Omega, \quad k \in \mathbb{N}_0. \quad (8.7)$$

Equation 8.7 may also be rewritten as

$$\frac{u^{(k+1)}(x) - u^{(k)}(x)}{\tau} + \mathcal{A}[u^{(k+1)}](x) = f(x, u^{(k)}(x)), \quad x \in \Omega, \quad k \in \mathbb{N}_0. \quad (8.8)$$

Introducing an artificial time t, making the displacement u time dependent, $u = u(x,t)$, and setting $u^{(k)}(x) = u(x, k\tau)$, where τ denotes a fixed time-step, eqn (8.7) may also be viewed as a semi-implicit scheme for the time-dependent partial differential equation

$$\partial_t u(x,t) + \mathcal{A}[u](x,t) = f(x, u(x,t)), \quad x \in \Omega. \quad (8.9)$$

A steady-state solution of eqn (8.9) also fulfills the necessary condition for a minimizer of Problem 8.1.

For a numerical treatment of either the fixed-point-type equation (8.6) or the time-dependent partial differential equation (8.9) two problems which are both related to discretization have to be solved: the computation of the force and the numerical solution of a partial differential equation.

8.3 Computing the forces

Since our main interest is the discussion of different smoothers, we focus on the distance measure $\mathcal{D} = \mathcal{D}^{\text{SSD}}$ already introduced in Section 6.1. The forces can be deduced from its Gâteaux derivative.

Theorem 8.1 *Let $d \in \mathbb{N}$ and $R, T \in \text{Img}(d)$, $T \in C^2(\mathbb{R}^d)$, $u, v : \mathbb{R}^d \to \mathbb{R}^d$, $\Omega :=]0,1[^d$. The Gâteaux derivative of $\mathcal{D}[R,T;u]$,*

$$\mathcal{D}[R,T;u] := \mathcal{D}[R,T;u] := \tfrac{1}{2}\,\|T_u - R\|_{L_2(\Omega)} \quad (8.10)$$

with respect to v is given by

$$d\mathcal{D}[R,T;u;v] = -\int_{\mathbb{R}^d} \langle f(x,u(x)), v(x)\rangle_{\mathbb{R}^d}\, dx,$$

where $f : \mathbb{R}^d \times \mathbb{R}^d \to \mathbb{R}^d$,

$$f(x,u(x)) := (R(x) - T_u(x))\nabla T_u(x). \quad (8.11)$$

Proof We take advantage of a Taylor expansion of $T(x - u(x) - hv(x))$ with respect to h at the expanding point $x - u(x)$,

$$T(x - u(x) - hv(x)) = T_u(x) - h \langle \nabla T_u(x), v(x) \rangle_{\mathbb{R}^d} + \mathcal{O}(h^2).$$

Thus,

$$\begin{aligned} & d\mathcal{D}[R, T; u; v] \\ & = \lim_{h \to 0} \frac{1}{h} (\mathcal{D}[R, T; u + hv] - \mathcal{D}[R, T; u]) \\ & = \lim_{h \to 0} \frac{1}{2h} \int_\Omega \left(T_u(x) - h \langle \nabla T_u(x), v(x) \rangle_{\mathbb{R}^d} + \mathcal{O}(h^2) - R(x) \right)^2 \\ & \quad - (T_u(x) - R(x))^2 dx \\ & = \int_\Omega \langle (R(x) - T_u(x)) \nabla T_u(x), v(x) \rangle_{\mathbb{R}^d} \, dx. \end{aligned}$$

□

Since the R and T are typically digital images, an interpolation scheme has to be used to compute $T_u(x)$ for non-integer values of $u(x)$. In our implementation, a bi- or tri-linear interpolation scheme (cf., Section 3.1.3) is used in order to keep the computation time reasonable.

8.4 General remarks on choosing the discretization

Due to the simplicity of the underlying domain $\Omega =]0,1[^d$, a finite difference scheme is used to solve the partial differential equation numerically. To this end, we make use of the grids introduced in Section 3.1.2.

For a function $g : \mathbb{R}^d \to \mathbb{R}$, we use the notation $\vec{g} := g(\vec{X}) := (g(x_j))_{j=1}^N \in \mathbb{R}^N$, and for a vector field $v : \mathbb{R}^d \to \mathbb{R}^d$, we set $\vec{v} := (\vec{v_1}^\top, \ldots, \vec{v_d}^\top)^\top \in \mathbb{R}^{dN}$.

From the Taylor expansion of $g : \mathbb{R}^d \to \mathbb{R}$, we get the standard finite difference approximation for second order derivatives,

$$\partial_{x_j x_j} g(x) = \frac{g(x + h_j e_j) - 2g(x) + g(x - h_j e_j)}{h_j^2} + \mathcal{O}(h_j^2), \tag{8.12}$$

$$\begin{aligned} \partial_{x_j x_k} g(x) = & \frac{1}{4 h_j h_k} (g(x + h_j e_j + h_k e_k) - g(x - h_j e_j + h_k e_k) \\ & - g(x + h_j e_j - h_k e_k) + g(x - h_j e_j - h_k e_k)) \\ & + \mathcal{O}(h_j^2 + h_k^2), \end{aligned} \tag{8.13}$$

where in the sequel we assume a fixed meshsize h for all spatial dimensions, i.e., $h_j = h$ for $j = 1, \ldots, d$. If this assumption is not fulfilled in an application, one

Algorithm 8.2 *General registration algorithm.*

Initialize $k = 0$, $\vec{X}^{(k)}$, and $\vec{U}^{(k)} = 0$.
For $k = 0, 1, 2, \ldots$
  compute force $\vec{F}^{(k)} = f(\vec{X}, \vec{U}^{(k)})$;
  solve partial differential equation, $A\vec{U}^{(k+1)} = \vec{F}^{(k)}$;
  if converged, stop, end;
end.

has either to embed the domain in a cube or to weight the coordinates reciprocal to the resolution. A detailed description is technical and omitted here.

We assume that a finite difference approximation of $\mathcal{A}[u]$ at any grid point x_j can be obtained using a convolution filter $S^{\mathcal{A}}$,

$$\mathcal{A}[g](x_j) \approx \sum_{\kappa \in \mathcal{N}(j) \cup j} S_{\kappa}^{\mathcal{A}} g(x_{\kappa}) =: (S^{\mathcal{A}} * g)(x_j)$$

where $\mathcal{N}(j)$ denotes a neighborhood of the grid point j and $S^{\mathcal{A}}$ denotes a filter connected with the partial differential operator $\mathcal{A}$ via its finite difference approximation. Of course, the particular boundary conditions attached to Problem 8.1 have to be incorporated into the discrete formulation. Practically, this can be done easily by a zero padding (Dirichlet), mirroring (Neumann), or adequate copying (periodic).

With the lexicographical ordering (cf., Definition 3.2), we can also deduce a matrix representation of the convolution with $S^{\mathcal{A}}$,

$$A \cdot \vec{g} := (S^{\mathcal{A}} * g)(\vec{X}).$$

Using for example the fixed-point type-iteration (8.6), we end up with an overall algorithm as summarized in Algorithm 8.2.

9

ELASTIC REGISTRATION

Broit (1981) was the first to study what he called *elastic registration*, a method based on a quantitative measure for the deformation, a quantitative measure for similarity between deformed images, and a procedure that uses the measures for obtaining an optimal mapping.

Broit's approach is based on

$$\mathcal{D}[R,T;u] = \mathcal{D}^{\mathrm{SSD}}[R,T;u] = \tfrac{1}{2}\left\|T_u - R\right\|_{L_2(\Omega)} \quad \text{and} \quad \mathcal{S}[u] := \mathcal{P}[u]$$

where $\mathcal{P}$ denotes the *linearized elastic potential* of the displacement u,

$$\mathcal{P}[u] = \int_\Omega \frac{\mu}{4} \sum_{j,k=1}^{d} (\partial_{x_j} u_k + \partial_{x_k} u_j)^2 + \frac{\lambda}{2}(\operatorname{div} u)^2 dx. \tag{9.1}$$

Here, λ and μ denote the so-called Lamé constants. For this particular choice of regularizer, the Euler–Lagrange equations are nothing more than the Navier–Lamé equations.

Navier–Lamé equations:

$$f = \mu\Delta u + (\lambda + \mu)\nabla \operatorname{div} u. \tag{9.2}$$

In the next section we give a physical motivation for this particular regularizer. The images are viewed as two different observations of an elastic body, one before and one after a *deformation*. The *displacement* of the elastic body is derived using a linear elasticity model. This development follows Broit (1981). In order to introduce different boundary conditions and for later reference, the eigenfunctions and eigenvalues of the Navier–Lamé operator are computed in Section 9.2. In Section 9.3 we show how the Navier–Lamé equations can be derived from the elastic potential using the general framework introduced in Chapter 8. This also provides a weak form of the Navier–Lamé equations.

Based on the general remarks on discretization (see Section 8.4) we derive a finite difference approximation of eqn (9.2). For ease of presentation, we focus on dimensions $d = 2, 3$. In Section 9.5 it is shown that a general class of matrices, including the discrete Navier–Lamé operator, can be diagonalized using fast Fourier transformation (FFT) techniques. The Moore–Penrose pseudo-inverse of the discrete Navier–Lamé operator is explicitly given for dimension $d = 2$ and traced back to the solution of small 3-by-3 linear systems for dimension $d = 3$. Finally, registration results for academic as well as real-life problems are presented.

9.1 Physical motivation

In this section we provide insight into the physical properties of an elastic body. The essential difference between a rigid and an elastic body is that the relation of particles is no longer fixed but can vary with respect to the elastic properties of the body; see Fig. 9.1.

Applying an external force to an elastic body results in a *deformation* or *strain* of the body. The strain is related to *tension* or *stress* of the body, and the shape of the body results from an equilibrium of outer forces and inner *stress*.

9.1.1 *Deformation of a body*

The deformation of a body or tissue is described by the time-dependent *transformation* $\varphi : \mathbb{R}^d \times \mathbb{R} \to \mathbb{R}^d$. Here, $\varphi(P,t)$ denotes the spatial position of a particle P at time t. For any fixed particle P, $x = \varphi(P,0)$ gives its original position, $\tilde{x} = \varphi(P,t_E)$ gives its position after deformation, and $\varphi(P,t)$, $0 \le t \le t_E$, gives the path of the particle P during deformation. For simplicity of notation, we also write $\varphi(x,t)$ for the position of the particle P at time t, where x is given by $x := \varphi(P,0)$. Note that we use the Lagrange coordinates to describe the physical behavior.

To begin with and for comparison reasons, we describe the transformation properties of a rigid body. Here, the distances and angles between particles remain fixed during transformation. Thus, the transformation is characterized by

$$\varphi(x,t_E) = Q \cdot \varphi(x,0) + b \quad \text{for all } x \in \mathbb{R}^d,$$

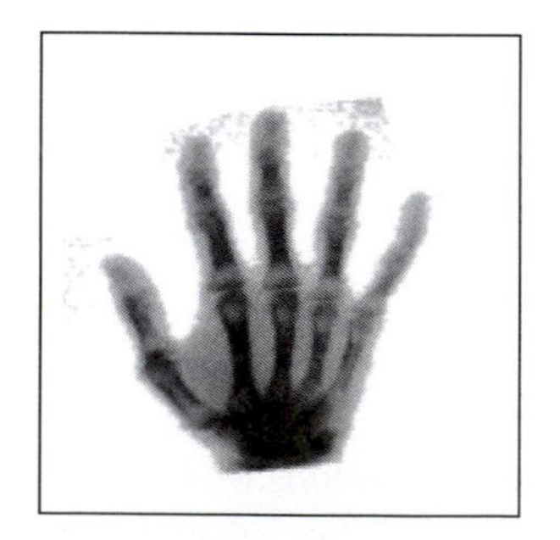
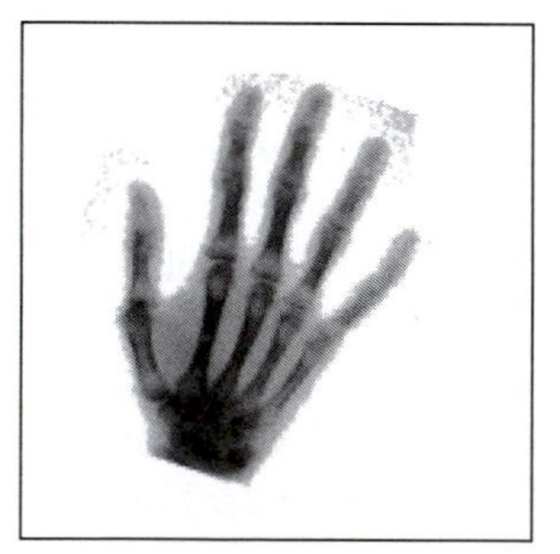
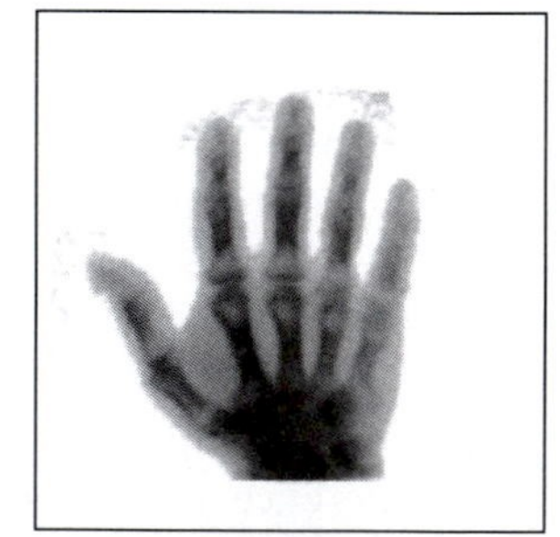

FIG. 9.1 LEFT: Image of a hand, MIDDLE: affine linear deformation of the hand, and RIGHT: elastic deformation of the hand.

where $b \in \mathbb{R}^d$ denotes a translation and $Q \in \mathbb{R}^{d\times d}$ denotes a rotation, i.e., $Q^\top Q = I_d$ and $\det Q = 1$. The distance $r := y - x$ of two particles x and y is mapped to $\tilde{r} := \tilde{y} - \tilde{x} = Q \cdot r$.

9.1.2 *The strain tensor*

For an elastic body, however, the transformation does not need to be rigid. For example, the distance between two particles can be expanded; see Fig. 9.2. Since we are interested in a linear elasticity model, only first order terms in $r = x - y$ will be considered. Thus, to a first order approximation, we have

$$\tilde{r} = \tilde{y} - \tilde{x} = \varphi(y, t_E) - \varphi(x, t_E) \approx \nabla\varphi(x, t_E)(y - x) = \nabla\varphi(x, t_E)r,$$

where

$$\nabla\varphi(x,t) := \begin{pmatrix} \partial_{x_1}\varphi_1 & \cdots & \partial_{x_d}\varphi_1 \\ \vdots & \ddots & \vdots \\ \partial_{x_1}\varphi_d & \cdots & \partial_{x_d}\varphi_d \end{pmatrix} \in \mathbb{R}^{d\times d}.$$

Introducing the *displacement* $u : \mathbb{R}^d \times \mathbb{R} \to \mathbb{R}^d$,

$$u(x,t) := \varphi(x,t) - x,$$

we find that $\nabla\varphi(x,t) = I_d + \nabla u(x,t)$.

Remark 9.1 Typically one is interested in a diffeomorphic transformation, i.e., the transformation is not allowed to introduce cracks and/or foldings of the tissue. Thus, it is worthwhile noticing that if $|\partial_{x_j} u_k| \ll 1$, we also have

$$\det(\nabla\varphi(x,t)) \approx 1 + \operatorname{div} u \neq 0$$

and hence φ is diffeomorphic.

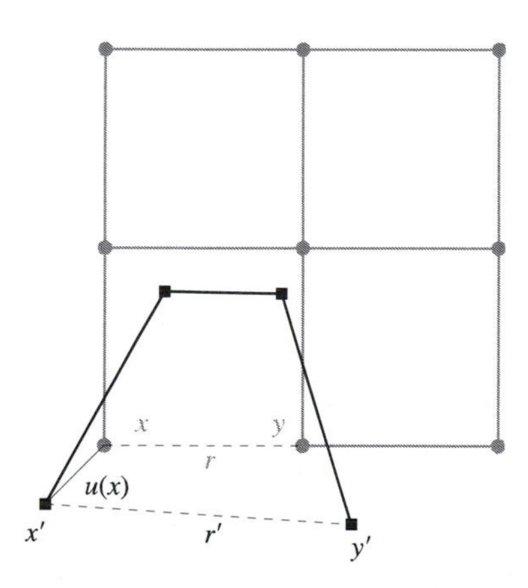

FIG. 9.2 Elastic body, particles before (dots) and after (squares) deformation.

Remark 9.2 For later purposes, we comment here on the material derivative. To this end, let P again be a fixed particle, $\varphi(P,t)$ the path followed by the particle P, and $x = \varphi(P,0)$ its initial position. Hence, $\varphi(P,t) = x + u(P,t)$ by definition of the displacement u. The velocity v of the particle is defined by

$$v(P,t) := \partial_t \varphi(P,t) = \partial_t u(P,t) = \frac{d}{dt} u(P,t),$$

where ∂_t and d/dt denote the partial and total derivative with respect to time, respectively. Note that P does not depend on the time.

In contrast to the particle focussing Lagrange coordinates we consider the position following Euler coordinates; see also Section 3.3.2. The displacement u and the velocity v can also be related to the spatial position,

$$\tilde{u}(\tilde{x},t) := u(P,t), \quad \tilde{v}(\tilde{x},t) := v(P,t),$$

where the particle and the position are connected by $\tilde{x} = \varphi(P,t)$.

From this point of view, we have

$$\tilde{v}(\tilde{x},t) = v(P,t) = \frac{d}{dt} u(P,t) = \frac{d}{dt} \tilde{u}(\tilde{x},t) = \frac{d}{dt} \tilde{u}(\varphi(P,t),t).$$

The material derivative is given by

$$\begin{aligned} \frac{d}{dt} \tilde{u}(\tilde{x},t) &= \nabla_x \tilde{u}(\varphi(P,t),t) \partial_t \varphi(P,t) + \partial_t \tilde{u}(\varphi(P,t),t) \\ &= \nabla_x \tilde{u}(\tilde{x},t) \tilde{v}(\tilde{x},t) + \partial_t \tilde{u}(\tilde{x},t), \end{aligned}$$

where $\nabla_x = (\partial_{x_1}, \dots, \partial_{x_d})^\top$, and the chain rule has been applied. Thus, the partial time derivative of $\tilde{u}$ can also be expressed by

$$\partial_t \tilde{u}(\tilde{x},t) = \tilde{v}(\tilde{x},t) - \nabla_x \tilde{u}(\tilde{x},t) \tilde{v}(\tilde{x},t).$$

In elasticity theory it is convenient to measure not just the displacement but its alterations. To this end, one introduces the *strain tensor*

$$V(x,t) := (\varepsilon_{j,k}(x,t))_{j,k} := \tfrac{1}{2}\big(\nabla u(x,t) + (\nabla u(x,t))^\top\big). \tag{9.3}$$

Note that, to a first order approximation,

$$[\nabla \varphi(x,t)]^\top \nabla \varphi(x,t) \approx I_d + \nabla u(x,t) + (\nabla u(x,t))^\top \approx (I_d + V(x,t))^2.$$

Thus, $V(x,t)$ might be viewed as a measure for the departure of rigidity of $\varphi(x,t)$. If in particular $\varphi(x) = Qx + b$ describes a rigid transformation, $V(x,t) = 0$. On the other hand, rotations cannot be measured with $V(x,t)$. From Remark 9.1, we have $\det(\nabla \varphi(x,t)) > 0$ and $\det(I_d + V(x,t)) > 0$. Thus,

$$Q(x,t) := (I_d + V(x,t))(\nabla \varphi(x,t))^{-1}$$

is a pure rotation. The transformation may be decomposed into rigid and non-rigid parts, where the non-rigid parts are given by

$$Q(x,t)\tilde{r} - r = V(x,t) \cdot r. \tag{9.4}$$

From eqn (9.4) we obtain that $V(x,t)$ describes the relative changes of lengths and angles introduced to the elastic body by the transformation φ. For dimension one this is obvious since $Q(x,t) = 1$ and $V(x,t) = \varepsilon_{1,1} = (\tilde{r} - r)/r$. Our next example illustrates this relation for dimension three.

Example 9.1 (Elastic body in $\mathbb{R}^3$) For the points

$$P_0 = \begin{pmatrix} x_0 \\ y_0 \\ z_0 \end{pmatrix}, \quad P_1 = \begin{pmatrix} x_0 + x \\ y_0 \\ z_0 \end{pmatrix}, \quad \text{and} \quad P_2 = \begin{pmatrix} x_0 \\ y_0 + y \\ z_0 \end{pmatrix},$$

we define $r_1 := \vec{P_0P_1} = xe_1$ and $r_2 := \vec{P_0P_2} = ye_2$. Considering only first order terms for the length of $\tilde{r}_1 = \varphi(P_1, t_E) - \varphi(P_0, t_E)$, we obtain

$$\begin{aligned} \|\tilde{r}_1\|^2_{\mathbb{R}^3} &\approx \|\nabla\varphi \cdot r_1\|^2_{\mathbb{R}^3} \\ &\approx r_1^\top (I_d + V)^2 r_1 \\ &\approx r_1^\top r_1 + 2r_1^\top V r_1 = x^2(1 + 2\varepsilon_{1,1}). \end{aligned}$$

Hence, utilizing a Taylor expansion, we get

$$\frac{\|\tilde{r}_1\|_{\mathbb{R}^3} - \|r_1\|_{\mathbb{R}^3}}{\|r_1\|_{\mathbb{R}^3}} \approx \sqrt{1 + 2\varepsilon_{1,1}} - 1 \approx \varepsilon_{1,1}.$$

Analogous considerations for $j = 2, 3$ show that $\varepsilon_{j,j}$ are the relative changes of lengths with respect to the direction spanned by e_j.

Computing for example

$$(\tilde{r}_1)^\top \tilde{r}_2 = r_1(I_d + V)^2 r_2 \approx r_1^\top r_2 + 2r_1^\top V r_2 = 2xy\varepsilon_{1,2},$$

we obtain for the change of angle

$$\begin{aligned} \delta = \angle(r_1, r_2) - \angle(\tilde{r}_1, \tilde{r}_2) &= \frac{\pi}{2} - \angle(\tilde{r}_1, \tilde{r}_2) \\ \approx \sin\delta = \frac{(\tilde{r}_1)^\top \tilde{r}_2}{\|\tilde{r}_1\|\,\|\tilde{r}_2\|} &\approx \frac{2xy\varepsilon_{12}}{x(1 + \varepsilon_{1,1})y(1 + \varepsilon_{2,2})} \approx 2\varepsilon_{12}, \end{aligned}$$

showing that $\varepsilon_{j,k}$ measures the relative change of angles of the non-rigid body.

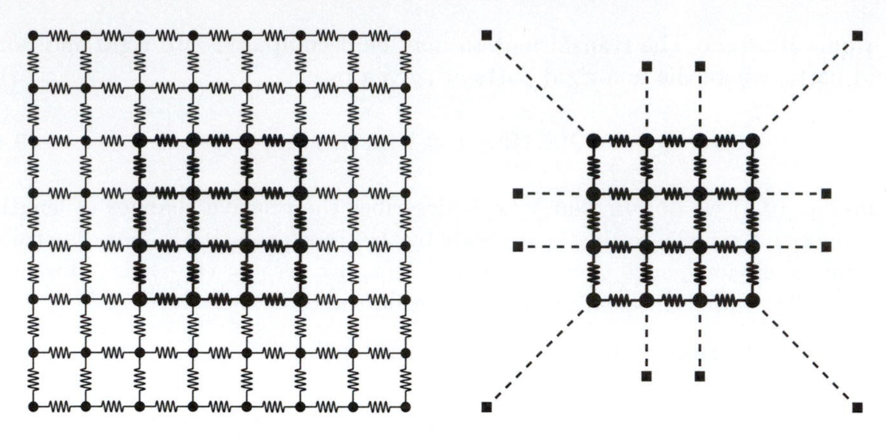

FIG. 9.3 LEFT: Two-dimensional elastic body with springs and part of the body (particles marked with dots); RIGHT: isolated part with strain at the boundary.

9.1.3 *The stress tensor*

The connectivity of an elastic body can be described in terms of *stress*. In Fig. 9.3, the particles of an elastic body idealized as regular grid points are shown. These particles are connected by small springs. The energy needed to maintain the shape of a particular part of the elastic body when isolated from the rest is expressed by the stress γ, where $\gamma : \mathbb{R}^d \times \mathbb{R} \times \mathbb{R}^d \to \mathbb{R}^d$. The stress might be viewed as a force field acting on the boundary $\partial\Omega$ of any part Ω of the body. The equations of equilibrium for the elastic body Ω are (cf., e.g., Gurtin (1983))

$$\int_\Omega f\,dx + \int_{\partial\Omega} \gamma\,dx = 0, \tag{9.5}$$

$$\int_\Omega x \times f\,dx + \int_{\partial\Omega} x \times \gamma\,dx = 0, \tag{9.6}$$

where f denotes an external force field and $\times$ denotes the vector cross-product.

For the elastic body under consideration we assume that there exist functions $\sigma_{j,k} : \mathbb{R}^d \times \mathbb{R} \to \mathbb{R}^d$, $j, k = 1, \ldots, d$, such that

$$\int_{\partial\Omega} \gamma_j\,dx = \int_{\partial\Omega} \sum_{k=1}^{d} \sigma_{j,k} n_k\,dx =: \int_{\partial\Omega} \langle \sigma_{j,:}, \vec{n} \rangle\,dx,$$

where $\vec{n} = (n_1, \ldots, n_d)$ denotes the outer normal on $\partial\Omega$. Using the Gaussian integral or divergence theorem (cf., e.g., Gurtin (1981, §2(5))), we obtain

$$\int_{\partial\Omega} \gamma_j\,dx = \int_\Omega \operatorname{div} \sigma_{j,:}\,dx, \quad j = 1, \ldots, d.$$

In other words, we can express the boundary integral on the left hand side as an integral over the body. The *stresses* $\sigma_{j,k}$ are collected in the so-called *stress tensor* Σ,

$$\Sigma(x,t) := (\sigma_{j,k})_{j,k=1,\ldots,d} \in \mathbb{R}^{d\times d}. \tag{9.7}$$

Physical considerations show that the stress tensor is symmetric, i.e., $\Sigma = \Sigma^\top$; cf., Example 9.2.

Starting with the equations of equilibrium and varying over all possible sub-bodies, we rediscover Newton's first law, the *balance of forces*,

$$f = -\operatorname{div}\Sigma,$$

and Newton's second law, the *balance of momentum*,

$$x \times f = -x \times \operatorname{div}\Sigma.$$

The next example motivates that for dimension three, the *stress tensor* is symmetric. To this end, we consider an arbitrary tetrahedron. Note that a general body might be covered by a set of tetrahedra.

Example 9.2 As a reference volume we consider a tetrahedron in $\mathbb{R}^3$ with edges

$$P_0 = \begin{pmatrix} x_0 \\ y_0 \\ z_0 \end{pmatrix}, \quad P_1 = \begin{pmatrix} x_0 + x \\ y_0 \\ z_0 \end{pmatrix}, \quad P_2 = \begin{pmatrix} x_0 \\ y_0 + y \\ z_0 \end{pmatrix}, \quad \text{and} \quad P_3 = \begin{pmatrix} x_0 \\ y_0 \\ z_0 + z \end{pmatrix};$$

cf., Fig. 9.4. The boundary surfaces are denoted by F_1, F_2, F_3, and F_4, where e_j is normal to F_j, $j = 1, 2, 3$, and $\vec{n} = (n_1, n_2, n_3)^\top$ is normal to F_4. Using

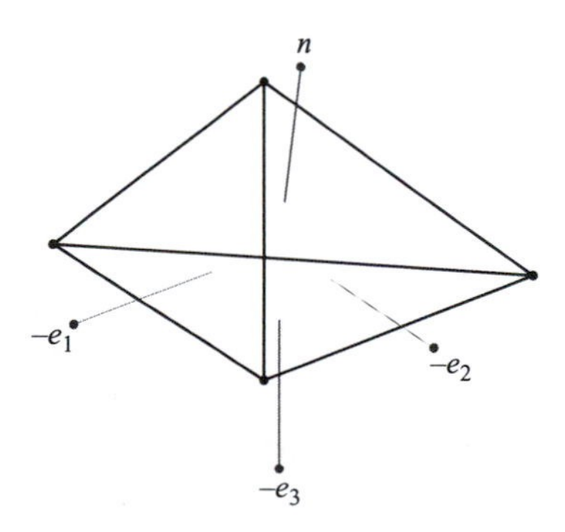

FIG. 9.4 Tetrahedron with outer normals $-e_1$, $-e_2$, $-e_3$, and $\vec{n}$.

elementary calculus for the cross-product $\times$, we obtain for the areas $|F_j|$,

$$\begin{aligned}2|F_4|\vec{n} &= \vec{P_1P_2} \times \vec{P_1P_3} \\ &= (\vec{P_0P_2} - \vec{P_0P_1}) \times (\vec{P_0P_3} - \vec{P_0P_1}) \\ &= \vec{P_0P_2} \times \vec{P_0P_3} - \vec{P_0P_1} \times \vec{P_0P_3} - \vec{P_0P_2} \times \vec{P_0P_1} \\ &= 2|F_1|e_1 + 2|F_2|e_2 + 2|F_3|e_3 = 2(|F_1|, |F_2|, |F_3|)^\top.\end{aligned}$$

For fixed numbers $a, b > 0$ and arbitrary $h > 0$ we obtain a family of congruent tetrahedra with $x = h$, $y = ah$, $z = bh$ having the same outer normals. For an arbitrarily chosen point $Q \in F_4$, we have $\vec{0Q} = \vec{0P_0} + \vec{P_0Q} = p + q$, where $p = \vec{0P_0}$ and $q = \vec{P_0Q}$. Since $Q \in F_4$, $\|q\|_{\mathbb{R}^3} \leq h^2(1 + a^2 + b^2)$, and with a $\theta \in [0, 1]$ we get for the jth component of the stress γ,

$$\begin{aligned}\gamma_j(Q, t, \vec{n}) &= \gamma_j(p + q, t, \vec{n}) \\ &= \gamma_j(p, t, \vec{n}) + \nabla\gamma_j(p + \theta q, t, \vec{n})^\top q \\ &= \gamma_j(p, t, \vec{n}) + \mathcal{O}(h).\end{aligned}$$

For the tetrahedron T with boundary ∂T, we thus find

$$\begin{aligned}\int_{\partial T} \gamma_j \, dx &= \int_{\partial T} \gamma_j(x, t, \vec{n}(x)) \, dx = \sum_{k=1}^{4} \int_{F_k} \gamma_j(x, t, \vec{n}(x)) \, dx \\ &= \int_{F_4} \gamma_j(x, t, \vec{n}) \, dx + \sum_{k=1}^{3} \int_{F_k} \gamma_j(x, t, -e_k) \, dx \\ &\approx \gamma_j(p + q, t, \vec{n})|F_4| + \sum_{k=1}^{3} \gamma_j(p, t, -e_k)|F_k| \\ &\approx |F_4| \left(\gamma_j(p, t, \vec{n}) + \mathcal{O}(h) + \sum_{k=1}^{3} \gamma_j(p, t, -e_k) n_k \right).\end{aligned}$$

With the abbreviation $\sigma_{j,k}(p, t) := \gamma_j(p, t, -e_k)$, we may also write

$$\lim_{h \to 0} \frac{\int_{\partial T} \gamma_j \, dx}{\int_{\partial T} dx} = \frac{1}{1 + n_1 + n_2 + n_3} \left(\gamma_j(p, t, n) - \sum_{k=1}^{3} \sigma_{j,k}(p, t) n_k \right). \tag{9.8}$$

Now, for a continuous force field f we have

$$\int_T f_j \, dx \leq C_1 \operatorname{Vol}(T) = C_1 \frac{1}{3} |F_3| \left\| \vec{P_0P_3} \right\|_{\mathbb{R}^d} = \frac{C_1}{3} |F_3| bh$$

and hence

$$\frac{\int_{\partial T} f_j \, dx}{\int_{\partial T} dx} = \frac{C_1}{3} \frac{|F_3|bh}{|F_1| + |F_2| + |F_3| + |F_4|} \leq C_2 bh \overset{h \to 0}{\longrightarrow} 0.$$

From eqn (9.8) we finally obtain

$$\gamma_j(p, t, \vec{n}) = \sum_{k=1}^{3} \sigma_{j,k}(p, t) n_k = \sigma_{j,:}^{\top} \vec{n}.$$

This derivation gives a linearized formulation of stress with respect to the normal vector $\vec{n}$. Moreover, starting with the equilibrium equation (9.5) and running over all possible bodies K we also find

$$-f_j = \operatorname{div} \sigma_{j,:} = \partial_{x_1} \sigma_{j,1} + \partial_{x_2} \sigma_{j,2} + \partial_{x_3} \sigma_{j,3},$$

which follows from eqn (9.5). Hence,

$$\begin{aligned}
(x \times f)_1 &= x_2 f_3 - x_3 f_2 \\
&= -x_2(\partial_{x_1} \sigma_{3,1} + \partial_{x_2} \sigma_{3,2} + \partial_{x_3} \sigma_{3,3}) \\
&\quad + x_3(\partial_{x_1} \sigma_{2,1} + \partial_{x_2} \sigma_{2,2} + \partial_{x_3} \sigma_{2,3}) \\
&= -\partial_{x_1}[x_2 \sigma_{3,1}] - \partial_{x_2}[x_2 \sigma_{3,2}] - \partial_{x_3}[x_2 \sigma_{3,3}] + \sigma_{3,2} \\
&\quad + \partial_{x_1}[x_3 \sigma_{2,1}] + \partial_{x_2}[x_3] \sigma_{2,2} + \partial_{x_3}[x_3 \sigma_{2,3}] - \sigma_{2,3} \\
&= \operatorname{div}(x_3 \sigma_{2,:} - x_2 \sigma_{3,:}) + \sigma_{3,2} - \sigma_{2,3},
\end{aligned}$$

and thus

$$\begin{aligned}
&\int_K (x \times f)_1 - \sigma_{3,2} + \sigma_{2,3} \, dx = \int_K \operatorname{div}(x_3 \sigma_{2,:} - x_2 \sigma_{3,:}) \, dx \\
&= \int_{\partial K} x_3 \sigma_{2,:}^{\top} n - x_2 \sigma_{3,:}^{\top} n \, dx = \int_{\partial K} x_3 \gamma_2 - x_2 \gamma_3 \, dx = -\int_{\partial K} (x \times \gamma)_1 \, dx.
\end{aligned}$$

Exploiting the equilibrium eqn (9.6) we obtain $\sigma_{3,2} - \sigma_{2,3} = 0$. Using similar considerations for the remaining components we obtain the symmetry of the stress tensor, i.e.,

$$\sigma_{j,k} = \sigma_{k,j} \quad \text{for all } j, k.$$

9.1.4 *Hooke's law*

We are now ready to link the different parts together. Since deformation is related to strain and the forces are related to stress, the missing link is the connection between strain and stress. This link is provided by Hooke's law. Hooke's law "*ut tensio sic vis*" claims that strain is a reaction of stress. Here we use a linear model and take into account the fact that without strain the body undergoes no stress. Moreover, we exploit the symmetry of the strain and stress tensors $V = (\varepsilon_{j,k})$ and $\Sigma = (\sigma_{j,k})$, respectively. For example, for dimension $d = 3$, these assumptions result in the linear model

$$\begin{pmatrix} \varepsilon_{1,1} \\ \varepsilon_{2,2} \\ \varepsilon_{3,3} \\ \varepsilon_{1,2} \\ \varepsilon_{1,3} \\ \varepsilon_{2,3} \end{pmatrix} = \begin{pmatrix} a_{1,1} & \cdots & a_{1,6} \\ \vdots & \ddots & \vdots \\ a_{6,1} & \cdots & a_{6,6} \end{pmatrix} \begin{pmatrix} \sigma_{1,1} \\ \sigma_{2,2} \\ \sigma_{3,3} \\ \sigma_{1,2} \\ \sigma_{1,3} \\ \sigma_{2,3} \end{pmatrix},$$

with 36 degrees of freedom. A considerable simplification occurs for isotropic elastic bodies. Here, the principal axes of strain and stress coincide. Thus, the strain and the stress tensors can be diagonalized simultaneously. Using an eigenvector system as a new reference coordinate system, and denoting the eigenvalues of V and Σ by ε_j and σ_j, respectively, where $j = 1, 2, \ldots, d$, the strain and stress tensors with respect to this coordinate system are given by

$$\hat{V} = \begin{pmatrix} \varepsilon_1 & & \\ & \ddots & \\ & & \varepsilon_d \end{pmatrix} \quad \text{and} \quad \hat{\Sigma} = \begin{pmatrix} \sigma_1 & & \\ & \ddots & \\ & & \sigma_d \end{pmatrix},$$

respectively. The linear model becomes

$$\begin{pmatrix} \varepsilon_1 \\ \vdots \\ \varepsilon_d \end{pmatrix} = \begin{pmatrix} b_{1,1} & \cdots & b_{1,d} \\ \vdots & \ddots & \vdots \\ b_{d,1} & \cdots & b_{d,d} \end{pmatrix} \begin{pmatrix} \sigma_1 \\ \vdots \\ \sigma_d \end{pmatrix}. \tag{9.9}$$

Moreover, since the reaction of the isotropic body is independent of the direction, the matrix has to circular, i.e., $b_{j,k} = b_{(k-j \bmod d)+1}$, and the non-diagonal elements have to be equal, i.e., $b_2 = \cdots = b_d$. Two parameters are thus sufficient to describe the properties of an elastic, isotropic body in any spatial dimension d. Introducing Young's modulus of elasticity E and Poisson's contraction ratio ν, we have $b_{j,j} = b_1 = E^{-1}$ and $b_{j,k} = b_k = -\nu E^{-1}$, for $j \neq k$.

The relation between stress and strain can now be rewritten with the material constants E and ν,

$$E\varepsilon_j = (1+\nu)\sigma_j - \nu \sum_{j=1}^{d} \sigma_j, \quad j = 1, \ldots, d,$$

and for the strain and stress tensors in the principal axes we have

$$E\hat{V} = (1+\nu)\hat{\Sigma} - \nu \, \mathrm{trace}(\hat{\Sigma}) I_d.$$

Changing back to the original basis we conclude

$$\begin{aligned} EV &= (1+\nu)\Sigma - \nu\,\mathrm{trace}(\Sigma) I_d, \\ E\,\mathrm{trace}(V) &= (1+\nu(1-d))\,\mathrm{trace}(\Sigma), \\ (1+\nu)\Sigma &= EV + \frac{E\nu}{1+\nu(1-d)}\,\mathrm{trace}(V) I_d. \end{aligned}$$

Introducing the Lamé constants

$$\mu := \frac{1}{2}\frac{E}{1+\nu}, \qquad \lambda := \frac{E\nu}{(1+\nu)(1+\nu(1-d))} \tag{9.10}$$

we can express stress in terms of strain and deformations with respect to general coordinates,

$$\begin{aligned} \sigma_{j,k} &= 2\mu\varepsilon_{j,k} + \lambda\,\mathrm{trace}(V)\delta_{j,k} \\ &= \mu(\partial_{x_j} u_k + \partial_{x_k} u_j) + \lambda\,\mathrm{div}\, u\,\delta_{j,k}. \end{aligned}$$

An elementary computation gives

$$\mathrm{div}\,\sigma_{j,:} = \sum_{k=1}^{d} \partial_{x_k}\sigma_{j,k} = \mu\Delta u_j + (\lambda+\mu)\partial_{x_j}\,\mathrm{div}\, u_j.$$

Summarizing these calculations we end up with the well-known Navier–Lamé equations

$$f = \mu\Delta u + (\lambda+\mu)\nabla\,\mathrm{div}\, u;$$

cf., eqn (9.2). Young's modulus and Poisson's ratio can be derived from the Lamé constants λ and μ,

$$E = \frac{2\mu(2\mu+d\lambda)}{2\mu+(d-1)\lambda}, \qquad \nu = \frac{\lambda}{2\mu+(d-1)\lambda}.$$

The constant μ is also called *shear modulus* in the literature. Note that $\lambda = \lambda_d$ depends on the spatial dimension d; cf., eqn (9.10). In particular for $d = 2$ and $d = 3$, we have

$$\lambda_2 = \frac{E\nu}{1-\nu} \quad \text{and} \quad \lambda_3 = \frac{E\nu}{1-2\nu}.$$

Our next example illustrates the role of μ and λ.

Example 9.3 We solve a discretized version of the Navier–Lamé equations on a rectangle for various values of the Lamé constants. For $\lambda = 0$, we have $E = 2\mu$ and $\nu = 0$. Thus, the body shows no contraction under deformation. On increasing the value of μ, the body becomes more rigid ($\mu = 10, 25, 50$ in our example, shown in Fig. 9.5).

For fixed μ, $E = E(\lambda)$ and $\nu = \nu(\lambda)$ are monotonically increasing functions with respect to λ: $E(0) = 2\mu$, $E(\infty) = 2\mu d/(d-1)$, and $E'(\lambda) = 4\mu^2/(2\mu + (d-1)\lambda)^2 > 0$, $\nu(0) = 0$, $\nu(\infty) = 1/(d-1)$, and $\nu'(\lambda) = 2\mu/(2\mu+(d-1)\lambda)^2 > 0$.

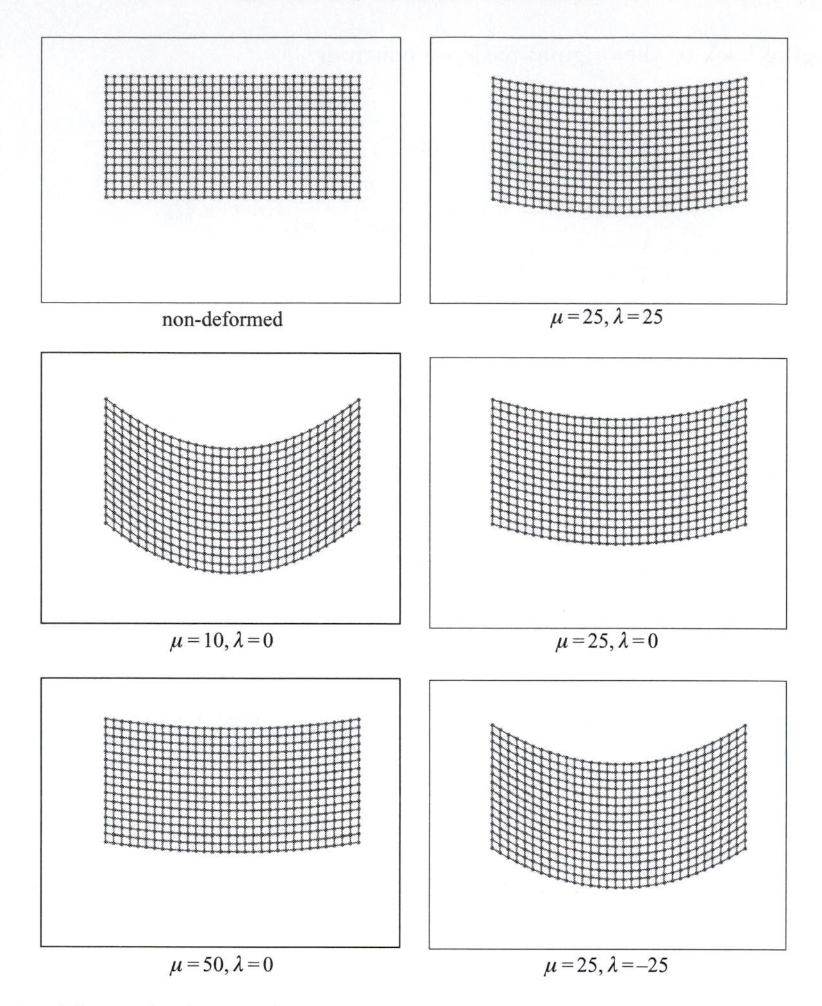

FIG. 9.5 Elastic bodies with various Lamé constants under deformation. Here, $f(x_1, x_2) = (0, -1)$ and the following boundary conditions are chosen: $u(L, x_2) = u(R, x_2) = 0$ on the left and right parts, $u_1(x_1, u) = u_1(x_1, l) = 0$, and $\partial_{x_1} u_2(x_1, u) = \partial_{x_1} u_2(x_1, l) = 0$ on the upper and lower parts.

On increasing λ, the body becomes more rigid ($\lambda = -25, 0, 25$ in our example). Note that $\lambda \leq 0$ is not physically possible, since E and ν are positive constants.

9.2 Eigenfunctions of the Navier–Lamé operator

For particular boundary conditions, eigenfunctions of the Navier–Lamé operator are known explicitly. This knowledge may serve as a basis for a numerical solution scheme for the Navier–Lamé equations. Expanding the displacement and the force with respect to the normalized eigenfunctions v_j,

$$u = \sum_{j=1}^{M} \alpha_j v_j \quad \text{and} \quad f = \sum_{j=1}^{M} \beta_j v_j,$$

respectively, we have

$$\mathcal{A}u = \sum_{j=1}^{M} \alpha_j \mathcal{A} v_j = \sum_{j=1}^{M} \alpha_j \lambda_j v_j = \sum_{j=1}^{M} \beta_j v_j = f,$$

showing that the solution u is given by $\alpha_j = 0$ if $\lambda_j = 0$ and $\alpha_j = \beta_j/\lambda_j$ otherwise. Here we assume the system to be consistent. However, this approach has two drawbacks. In a numerical implementation the length M of the expansion has to be finite and one needs to compute the coefficients $\beta_j = \langle f, v_j \rangle$ numerically.

Bro-Nielsen & Gramkow (1996) used a modification of this technique. They worked with basis functions v_j, where $v_j(x) = v_0(x_j - x)$, and computed $\tilde{\beta}_j = \langle v_j, f \rangle = (v_0 * f)(x_j)$. This leads to a considerable improvement with respect to the numerical complexity of the method. However, a statement concerning the quality of this solution is missing.

We now give explicit formulas for the eigenvalues and eigenfunctions of the Navier–Lamé operator $\mathcal{A}$,

$$\mathcal{A}[u] := \mu \Delta u + (\lambda + \mu) \nabla \operatorname{div} u, \tag{9.11}$$

i.e., we are looking for a function $v : \Omega \subset \mathbb{R}^d \to \mathbb{R}^d$ and a constant κ, satisfying

$$\mathcal{A}[v] = \kappa v.$$

To facilitate the discussion, we assume $d = 3$ and $\Omega =]0,1[^3$ throughout this section.

From the variational formulation we have implicit boundary conditions; cf., Theorem 9.4. In this section, however, we impose explicit second order boundary conditions on u. These boundary conditions result in an alteration of the partial differential operator.

Three different types of boundary conditions are discussed: *sliding*, *bending*, and *periodic*. With sliding boundary conditions the boundary is mapped onto the boundary but boundary points are allowed to slide on the boundary. Using bending boundary conditions, the corners of the domain are fixed but the sides are allowed to bend in- and outward. Finally, for the periodic boundary conditions, the domain Ω is mapped onto a torus.

Note that the results may also be found in the literature; cf., e.g., Christensen (1994). However, the proofs presented here appear to be new.

9.2.1 *Sliding boundary conditions*

The sliding boundary conditions considered here are defined by

$$
\begin{aligned}
u_1(z,x_2,x_3) &= u_2(x_1,z,x_3) = u_3(x_1,x_2,z) = 0,\\
\partial_{x_2} u_1(x_1,z,x_3) &= \partial_{x_3} u_1(x_1,x_2,z) = \partial_{x_1} u_2(z,x_2,x_3) = 0,\\
\partial_{x_3} u_2(x_1,x_2,z) &= \partial_{x_1} u_3(z,x_2,x_3) = \partial_{x_2} u_3(x_1,z,x_3) = 0,
\end{aligned}
$$

where $z = 0,1$ and $x_j \in [0,1]$.

Note that for this particular choice of boundary conditions the transformation φ, with $\varphi(x) = x + u(x)$, maps the boundary $\partial\Omega$ onto $\partial\Omega$. Since a particle on a side of the unit cube may slide along the side it belongs to, these boundary conditions are also called *sliding boundary conditions.*

Theorem 9.1 *Let* $d = 3$, $\Omega =]0,1[^3$. *For* $a,b,c \in \mathbb{N}_0$ *with* $a^2+b^2+c^2 \neq 0$, *let* $p_1 = (a,b,c)^\top$, $p_2 \perp p_1$, $p_2 \neq 0$, *and* $p_3 := p_1 \times p_2$. *Moreover, let*

$$
w_{abc}(x_1,x_2,x_3) := \begin{pmatrix} \sin(a\pi x_1)\cos(b\pi x_2)\cos(c\pi x_3)\\ \cos(a\pi x_1)\sin(b\pi x_2)\cos(c\pi x_3)\\ \cos(a\pi x_1)\cos(b\pi x_2)\sin(c\pi x_3)\end{pmatrix}.
$$

Then

$$
v_{abc;j} := \operatorname{diag}(p_j) w_{abc}
$$

is an eigenfunction of the partial differential operator $\mathcal{A}$ *(cf., eqn (9.11)) to the eigenvalue*

$$
\kappa_{abc;j} = -\pi^2(a^2+b^2+c^2)\begin{cases} (2\mu+\lambda), & j = 1\\ \mu, & j = 2,3,\end{cases}
$$

with respect to the sliding boundary conditions.

Proof For fixed a, b, and c, we define $w := w_{abc}$. Hence

$$
\begin{aligned}
\Delta w &= -\pi^2(a^2+b^2+c^2)w,\\
\operatorname{div} w &= \pi(a+b+c)\cos(a\pi x_1)\cos(b\pi x_2)\cos(c\pi x_3),\\
\nabla \operatorname{div} w &= -\pi^2(a+b+c)\operatorname{diag}(a,b,c)w.
\end{aligned}
$$

It is easy to verify that w and hence $v_{abc;j}$ fulfill the boundary conditions. Exploiting the eigenfunction representation

$$
\begin{aligned}
\mathcal{A}[\operatorname{diag}(\alpha,\beta,\gamma)w] &= \mu\operatorname{diag}(\alpha,\beta,\gamma)\Delta w + (\lambda+\mu)\nabla\operatorname{div}[\operatorname{diag}(\alpha,\beta,\gamma)w]\\
&= -\mu\pi^2(a^2+b^2+c^2)\operatorname{diag}(\alpha,\beta,\gamma)w\\
&\quad -\pi^2(\lambda+\mu)(a\alpha+b\beta+c\gamma)\operatorname{diag}(a,b,c)w\\
&= \kappa\cdot\operatorname{diag}(\alpha,\beta,\gamma)w,
\end{aligned}
$$

we find three solutions, either $\alpha = a$, $\beta = b$, $\gamma = c$, and hence $\kappa = -\pi^2(a^2+b^2+c^2)(2\mu+\lambda)$, or $a\alpha+b\beta+c\gamma = 0$ and hence $\kappa = -\pi^2(a^2+b^2+c^2)\mu$. Thus, for any triple (a, b, c) we can find three associated eigenfunctions $v_{abc;j} = \operatorname{diag}(p_j)w$. □

9.2.2 *Bending boundary conditions*

The bending boundary conditions considered here are defined by

$$\begin{aligned} u_1(x_1, z, x_3) = u_1(x_1, x_2, z) = u_2(z, x_2, x_3) = 0, \\ u_2(x_1, x_2, z) = u_3(z, x_2, x_3) = u_3(x_1, z, x_3) = 0, \\ \partial_{x_1} u_1(z, x_2, x_3) = \partial_{x_2} u_2(x_1, z, x_3) = \partial_{x_3} u_3(x_1, x_2, z) = 0, \end{aligned}$$

where $z = 0, 1$ and $x_j \in [0, 1]$. Here, the transformation might move the boundary of the unit cube.

Theorem 9.2 *Let $d = 3$, $\Omega =]0, 1[^3$. For $a, b, c \in \mathbb{N}_0$ with $a^2 + b^2 + c^2 \neq 0$, let $p_1 = (a, b, c)^\top$, $p_2 \perp p_1$, $p_2 \neq 0$, and $p_3 := p_1 \times p_2$. Moreover, let*

$$w_{abc}(x_1, x_2, x_3) := \begin{pmatrix} \cos(a\pi x_1)\sin(b\pi x_2)\sin(c\pi x_3) \\ \sin(a\pi x_1)\cos(b\pi x_2)\sin(c\pi x_3) \\ \sin(a\pi x_1)\sin(b\pi x_2)\cos(c\pi x_3) \end{pmatrix}.$$

Then

$$v_{abc;j} := \operatorname{diag}(p_j) w_{abc}$$

is an eigenfunction of the partial differential operator $\mathcal{A}$ (cf., eqn (9.11)) to the eigenvalue

$$\kappa_{abc;j} = -\pi^2(a^2 + b^2 + c^2) \begin{cases} (2\mu + \lambda), & j = 1 \\ \mu, & j = 2, 3, \end{cases}$$

with respect to the bending boundary conditions.

Proof The proof is along the same lines as the one for Theorem 9.1. □

9.2.3 *Periodic boundary conditions*

The periodic conditions considered here are defined by

$$\begin{aligned} u_1(0, x_2, x_3) = u_1(1, x_2, x_3), \\ u_2(x_1, 0, x_3) = u_2(x_1, 1, x_3), \\ u_3(x_1, x_2, 0) = u_3(x_1, x_2, 1). \end{aligned}$$

Theorem 9.3 *Let* $d = 3$, $\Omega =]0,1[^3$. *For* $a, b, c \in \mathbb{N}_0$ *with* $a^2 + b^2 + c^2 \neq 0$, *let* $p_1 = (a, b, c)^\top$, $p_2 \perp p_1$, $p_2 \neq 0$, *and* $p_3 := p_1 \times p_2$. *Moreover, for* $r, s, t = 0, 1$, *let*

$$w_{abc}^{rst}(x) := \begin{pmatrix} -\sin(a\pi x_1 + r\pi/2)\sin(b\pi x_2 + s\pi/2)\sin(c\pi x_3 + t\pi/2) \\ \cos(a\pi x_1 + r\pi/2)\cos(b\pi x_2 + s\pi/2)\sin(c\pi x_3 + t\pi/2) \\ \cos(a\pi x_1 + r\pi/2)\sin(b\pi x_2 + s\pi/2)\cos(c\pi x_3 + t\pi/2) \end{pmatrix}.$$

Then

$$v_{abc;j}^{rst} := \operatorname{diag}(p_j) w_{abc}^{rst}$$

is an eigenfunction of the partial differential operator $\mathcal{A}$ *(cf., eqn (9.11)) to the eigenvalue*

$$\kappa_{abc;j} = -\pi^2(a^2 + b^2 + c^2) \begin{cases} (2\mu + \lambda), & j = 1 \\ \mu, & j = 2, 3, \end{cases}$$

with respect to the periodic boundary conditions.

Proof By construction w_{abc}^{rst} and thus $v_{abc;j}^{rst}$ are periodic. We only prove the case $r = s = t = 0$. The proofs of the other cases are similar. Putting

$$w(x) = \begin{pmatrix} -\sin(a\pi x_1)\sin(b\pi x_2)\sin(c\pi x_3) \\ \cos(a\pi x_1)\cos(b\pi x_2)\sin(c\pi x_3) \\ \cos(a\pi x_1)\sin(b\pi x_2)\cos(c\pi x_3) \end{pmatrix}$$

we compute

$$\begin{aligned} \Delta w &= -\pi^2(a^2 + b^2 + c^2) w \\ \operatorname{div} w &= -\pi(a + b + c)\cos(a\pi x_1)\sin(b\pi x_2)\sin(c\pi x_3) \\ \nabla \operatorname{div} w &= -\pi^2(a + b + c)\operatorname{diag}(a, b, c) w. \end{aligned}$$

The rest of the proof is identical to that of Theorem 9.1. □

9.3 Variational formulation of the elasticity problem

We are now ready to relate the framework introduced in the previous chapter to elastic registration. We provide a variational formulation of the elasticity problem. The required displacement is characterized as a minimizer of a certain functional $\mathcal{J}$. The functional consists of two terms. The first term $a[u, u]$ is a bi-linear form and serves as a regularizer, modeling the material properties and the inner forces, and the second term $b[u]$ reflects the outer forces.

Let $\Omega =]0,1[^d \subset \mathbb{R}^d$ and $V[u] = (\varepsilon_{j,k}[u])_{j,k} \in \mathbb{R}^{d \times d}$ be the strain tensor, with $\varepsilon_{jk}[u] = \frac{1}{2}(\partial_{x_j} u_k + \partial_{x_k} u_j)$; cf., eqn (9.3). We make use of the positive semi-definite bi-linear form

$$
\begin{aligned}
a[u,v] &:= \int_\Omega 2\mu\,\mathrm{trace}(V[u]^\top V[v]) \\
&\qquad + \lambda\,\mathrm{trace}(V[u])\,\mathrm{trace}(V[v])\,dx \\
&= \int_\Omega 2\mu \sum_{j,k=1}^{d} \varepsilon_{j,k}[u]\varepsilon_{j,k}[v] + \lambda\,\mathrm{div}\,u\cdot\mathrm{div}\,v\,dx \\
&= \int_\Omega \frac{\mu}{2} \sum_{j,k=1}^{d} (\partial_{x_j}u_k + \partial_{x_k}u_j)(\partial_{x_j}v_k + \partial_{x_k}v_j) \\
&\qquad + \lambda\,\mathrm{div}\,u\cdot\mathrm{div}\,v\,dx \\
&= \int_\Omega \mu \sum_{k=1}^{d} \langle \nabla u_k + \partial_{x_k}u, \nabla v_k\rangle_{\mathbb{R}^d} + \lambda\,\mathrm{div}\,u\,\mathrm{div}\,v\,dx \qquad (9.12)
\end{aligned}
$$

and the linear form

$$
b[v] := \int_\Omega \langle f, v\rangle_{\mathbb{R}^d}\,dx. \qquad (9.13)
$$

A minimizer of the functional $\mathcal{J}$,

$$
\mathcal{J}[u] := \tfrac{1}{2}a[u,u] + b[u], \qquad (9.14)
$$

with Dirichlet, Neumann, or periodic boundary conditions for u, is characterized by $d\mathcal{J}[u;v] = 0$ for all appropriate v.

The following theorem may be phrased for less smooth functions u and v as well. However, as we intend to solve the resulting partial differential equations by means of a finite difference scheme, the considered smoothness conditions are what we want.

Theorem 9.4 *Let $\mathcal{J}$ be defined by eqn (9.14), a be defined by eqn (9.12), and b be defined by eqn (9.13), respectively. Moreover, let $u \in (C^2(\mathbb{R}^d))^d$. For the perturbation $v \in (C^2(\mathbb{R}^d))^d$, the Gâteaux derivative of $\mathcal{J}$ is given by*

$$
d\mathcal{J}[u;v] = \int_\Omega \langle f - \mu\Delta u - (\mu+\lambda)\nabla\,\mathrm{div}\,u, v\rangle_{\mathbb{R}^d}\,dx.
$$

Proof A computation gives

$$
\begin{aligned}
d\mathcal{J}[u;v] &= \lim_{h\to 0}\frac{1}{h}(\mathcal{J}[u+hv]-\mathcal{J}[u]) = a[u,v]+b[v] \\
&= \int_\Omega \sum_{k=1}^{d} \mu\,\langle \nabla u_k + \partial_{x_k} u, \nabla v_k\rangle_{\mathbb{R}^d} + \lambda \operatorname{div} u \operatorname{div} v + \langle f, v\rangle_{\mathbb{R}^d}\,dx \\
&= \int_{\partial\Omega} \mu \sum_{k=1}^{d} v_k\,\langle \nabla u_k + \partial_{x_k} u, \vec{n}\rangle_{\mathbb{R}^d} + \lambda \operatorname{div} u\,\langle v, \vec{n}\rangle_{\mathbb{R}^d}\,dx \\
&\quad + \int_\Omega \langle f - \mu\Delta u - (\mu+\lambda)\nabla \operatorname{div} u, v\rangle_{\mathbb{R}^d}\,dx,
\end{aligned}
$$

where $\vec{n}$ denotes the outer normal vector on $\partial\Omega$. Exploiting the implicit boundary conditions $\operatorname{div} u = \langle \nabla u_k + \partial_{x_k} u, \vec{n}\rangle_{\mathbb{R}^d} = 0$ on $\partial\Omega$, the boundary integral vanishes, which completes the proof. □

Note that Theorem 9.4 provides a weak formulation of the Navier–Lamé equations (9.2). In image registration, however, where one is typically interested in very smooth deformations, the weak formulation of the problem provides no particular advantage in comparison to a strong formulation.

With this particular bi-linear form, we can express the elastic potential of a displacement u for any dimension d.

Definition 9.1 *Let* $d \in \mathbb{N}$, $u : \mathbb{R}^d \to \mathbb{R}^d$, $u_j \in C^2(\mathbb{R}^d)$, $j = 1, \ldots, d$. *The* elastic potential $\mathcal{P}$ *of the displacement* u *is defined by*

$$
\begin{aligned}
\mathcal{P}[u] &:= \frac{1}{2} a[u,u] \\
&= \int_\Omega \mu \operatorname{trace}(V[u]^\top V[u]) + \frac{\lambda}{2}\operatorname{trace}(V[u])^2\,dx \\
&= \int_\Omega \frac{\mu}{4} \sum_{j,k=1}^{d} (\partial_{x_j} u_k + \partial_{x_k} u_j)^2 + \frac{\lambda}{2}(\operatorname{div} u)^2\,dx.
\end{aligned}
$$

We note that the elastic potential is invariant under rigid transformations. If considered with additional explicit homogeneous Dirichlet boundary conditions, the bi-linear form a is positive definite.

For the important cases, dimension $d = 2$ and $d = 3$, we give explicit formulas for the elastic potential $\mathcal{P}_d$,

$$\mathcal{P}_2[u] = \int_{\mathbb{R}^2} \frac{\lambda}{2}(\partial_{x_1} u_1 + \partial_{x_2} u_2)^2 + \mu\Big\{(\partial_{x_1} u_1)^2 + (\partial_{x_2} u_2)^2 + \frac{1}{2}(\partial_{x_1} u_2 + \partial_{x_2} u_1)^2\Big\}\, dx,$$

$$\mathcal{P}_3[u] = \int_{\mathbb{R}^3} \frac{\lambda}{2}(\partial_{x_1} u_1 + \partial_{x_2} u_2 + \partial_{x_3} u_3)^2 + \mu\Big\{(\partial_{x_1} u_1)^2 + (\partial_{x_2} u_2)^2 + (\partial_{x_3} u_3)^2\Big\} + \frac{\mu}{2}\Big\{(\partial_{x_1} u_2 + \partial_{x_2} u_1)^2 + (\partial_{x_1} u_3 + \partial_{x_3} u_1)^2 + (\partial_{x_2} u_3 + \partial_{x_3} u_2)^2\Big\}\, dx.$$

9.4 Finite difference approximation of the Navier–Lamé equations

We investigate numerical schemes for the minimization of $\mathcal{J}$,

$$\mathcal{J}[u] = \mathcal{D}[R, T; u] + \mathcal{P}[u],$$

where $\mathcal{D}$ is given by eqn (8.10) and $\mathcal{P}$ is the elastic potential; cf., Definition 9.1.

Our numerical treatment is based on the Euler–Lagrange equations, which coincide with the Navier–Lamé equations (9.2) for this particular regularizer. In particular, we discuss a scheme based on the fixed-point iteration eqn (8.6). Moreover, the scheme is based on a finite difference approximation for semi-linear partial differential equations. This finite difference approximation with respect to periodic boundary conditions (cf., Section 9.2.3) is explicitly given for dimension $d = 2, 3$.

The main point is that due to the choices of the particular boundary conditions and discretization, we finally end up with a high-dimensional linear system of equations, where the matrix is highly structured. This structure enables us to apply fast Fourier transformation techniques to explicitly invert the matrix in a complexity $\mathcal{O}(N \log N)$, where N is the number of image voxels.

9.4.1 *The two-dimensional case*

For the Navier–Lamé equations (9.2) and dimension $d = 2$ we have

$$\mathcal{A}[u] = \mu\Delta u + (\lambda + \mu)\nabla \operatorname{div} u = \begin{pmatrix} (\lambda + 2\mu)\partial_{x_1x_1} u_1 + \mu\partial_{x_2x_2} u_1 + (\lambda + \mu)\partial_{x_1x_2} u_2 \\ (\lambda + \mu)\partial_{x_1x_2} u_1 + \mu\partial_{x_1x_1} u_2 + (\lambda + 2\mu)\partial_{x_2x_2} u_2 \end{pmatrix}.$$

Hence, with

$$\begin{aligned}\mathcal{A}^{1,1}[u_1] &:= (\lambda+2\mu)\partial_{x_1x_1}u_1 + \mu\partial_{x_2x_2}u_1,\\ \mathcal{A}^{1,2}[u_2] &:= (\lambda+\mu)\partial_{x_1x_2}u_2,\\ \mathcal{A}^{2,1}[u_1] &:= (\lambda+\mu)\partial_{x_1x_2}u_1,\\ \mathcal{A}^{2,2}[u_2] &:= \mu\partial_{x_1x_1}u_2 + (\lambda+2\mu)\partial_{x_2x_2}u_2,\end{aligned}$$

and, using eqns (8.12) and (8.13),

$$\left.\begin{aligned} S^{1,1} &:= (S^{2,2})^\top = \begin{pmatrix} 0 & (\lambda+2\mu) & 0\\ \mu & -2(\lambda+3\mu) & \mu\\ 0 & (\lambda+2\mu) & 0\end{pmatrix},\\ S^{1,2} &:= (S^{2,1})^\top := \frac{\lambda+\mu}{4}\begin{pmatrix} 1 & 0 & -1\\ 0 & 0 & 0\\ -1 & 0 & 1\end{pmatrix},\end{aligned}\right\} \tag{9.15}$$

we have

$$\begin{aligned}\mathcal{A}[u](X) &= \begin{pmatrix}\mathcal{A}^{1,1}[u_1](X) + \mathcal{A}^{1,2}[u_2](X)\\ \mathcal{A}^{2,1}[u_1](X) + \mathcal{A}^{2,2}[u_2](X)\end{pmatrix}\\ &\approx \begin{pmatrix} S^{1,1} * u_1(X) + S^{1,2} * u_2(X)\\ S^{2,1} * u_1(X) + S^{2,2} * u_2(X)\end{pmatrix}\end{aligned}$$

or

$$\mathcal{A}[u](\vec{X}) \approx \begin{pmatrix} A^{1,1}u_1(\vec{X}) + A^{1,2}u_2(\vec{X})\\ A^{2,1}u_1(\vec{X}) + A^{2,2}u_2(\vec{X})\end{pmatrix} =: A\vec{u}.$$

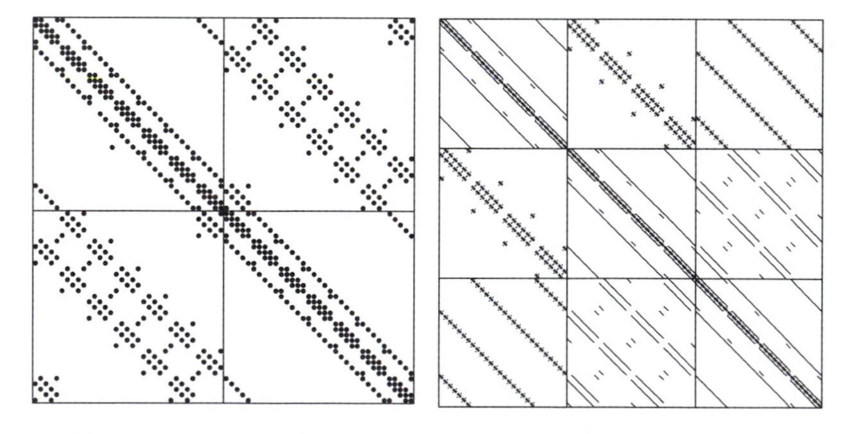

FIG. 9.6 Non-zero pattern of the matrix A for $d = 2$ (LEFT) and $d = 3$ (RIGHT), $n_1 = 5$, $n_2 = 7$, $n_3 = 4$.

Note that different boundary conditions result in different convolution schemes; see also Section 8.4. A typical non-zero pattern of the matrix A is shown in Fig. 9.6.

9.4.2 *The three-dimensional case*

For the Navier–Lamé equations (9.2) and dimension $d = 3$ we have

$$\begin{aligned}\mathcal{A}[u] &= \mu\Delta u + (\lambda+\mu)\nabla \operatorname{div} u \\ &= \begin{pmatrix}(\lambda+2\mu)\partial_{x_1x_1}u_1 + \mu\partial_{x_2x_2}u_1 + \mu\partial_{x_3x_3}u_1 \\ \mu\partial_{x_1x_1}u_2 + (\lambda+2\mu)\partial_{x_2x_2}u_2 + \mu\partial_{x_3x_3}u_2 \\ \mu\partial_{x_1x_1}u_3 + \mu\partial_{x_2x_2}u_3 + (\lambda+2\mu)\partial_{x_3x_3}u_3\end{pmatrix} \\ &\quad + (\lambda+\mu)\begin{pmatrix}\partial_{x_1x_2}u_2 + \partial_{x_1x_3}u_3 \\ \partial_{x_1x_2}u_1 + \partial_{x_2x_3}u_3 \\ \partial_{x_1x_3}u_1 + \partial_{x_2x_3}u_2\end{pmatrix}.\end{aligned}$$

Hence, with

$$\begin{aligned}\mathcal{A}^{1,1}[u_1] &:= (\lambda+2\mu)\partial_{x_1x_1}u_1 + \mu\partial_{x_2x_2}u_1 + \mu\partial_{x_3x_3}u_1, \\ \mathcal{A}^{1,2}[u_2] &:= (\lambda+\mu)\partial_{x_1x_2}u_2, \\ \mathcal{A}^{1,3}[u_3] &:= (\lambda+\mu)\partial_{x_1x_3}u_3, \\ \mathcal{A}^{2,1}[u_1] &:= (\lambda+\mu)\partial_{x_1x_2}u_1 \\ \mathcal{A}^{2,2}[u_2] &:= \mu\partial_{x_1x_1}u_2 + (\lambda+2\mu)\partial_{x_2x_2}u_2 + \mu\partial_{x_3x_3}u_2, \\ \mathcal{A}^{2,3}[u_3] &:= (\lambda+\mu)\partial_{x_2x_3}u_3, \\ \mathcal{A}^{3,1}[u_1] &:= (\lambda+\mu)\partial_{x_1x_3}u_1 \\ \mathcal{A}^{3,2}[u_2] &:= (\lambda+\mu)\partial_{x_2x_3}u_2 \\ \mathcal{A}^{3,3}[u_2] &:= \mu\partial_{x_1x_1}u_3 + \mu\partial_{x_2x_2}u_3 + (\lambda+2\mu)\partial_{x_3x_3}u_3,\end{aligned}$$

and with the matrix stencils summarized in Table 9.1, where eqns (8.12) and (8.13) are used, we have

$$\begin{aligned}\mathcal{A}[u](X) &= \begin{pmatrix}\mathcal{A}^{1,1}[u_1](X) + \mathcal{A}^{1,2}[u_2](X) + \mathcal{A}^{1,3}[u_3](X) \\ \mathcal{A}^{2,1}[u_1](X) + \mathcal{A}^{2,2}[u_2](X) + \mathcal{A}^{2,3}[u_3](X) \\ \mathcal{A}^{3,1}[u_1](X) + \mathcal{A}^{3,2}[u_2](X) + \mathcal{A}^{3,3}[u_3](X)\end{pmatrix} \\ &\approx \begin{pmatrix}S^{1,1} * u_1(X) + S^{1,2} * u_2(X) + S^{1,3} * u_3(X) \\ S^{2,1} * u_1(X) + S^{2,2} * u_2(X) + S^{2,3} * u_3(X) \\ S^{3,1} * u_1(X) + S^{3,2} * u_2(X) + S^{3,3} * u_3(X)\end{pmatrix}\end{aligned}$$

or

$$\mathcal{A}[u](\vec{X}) \approx \begin{pmatrix}A^{1,1}u_1(\vec{X}) + A^{1,2}u_2(\vec{X}) + A^{1,3}u_3(\vec{X}) \\ A^{2,1}u_1(\vec{X}) + A^{2,2}u_2(\vec{X}) + A^{2,3}u_3(\vec{X}) \\ A^{3,1}u_1(\vec{X}) + A^{3,2}u_2(\vec{X}) + A^{3,3}u_3(\vec{X})\end{pmatrix} =: A\vec{u}.$$

Table 9.1 *Elastic registration: matrix stencils for dimension $d = 3$.*

$$S^{1,1}_{j,k,\ell} = \alpha_{j,k,\ell}, \quad S^{2,2}_{j,k,\ell} = \alpha_{k,\ell,j}, \quad S^{3,3}_{j,k,\ell} = \alpha_{\ell,j,k},$$

$$\alpha_{j,k,\ell} = \begin{cases} -2(\lambda + 4\mu), & j = k = \ell = 2, \\ \lambda + 2\mu, & j = 1,3, \ k = \ell = 2, \\ \mu, & j = 2, \ k = 2, \ \ell = 1,3, \\ \mu, & j = 2, \ \ell = 2, \ k = 1,3, \\ 0, & \text{else}, \end{cases}$$

$$S^{1,2}_{j,k,\ell} = S^{2,1}_{j,k,\ell} = q(\ell, j, k),$$

$$S^{1,3}_{j,k,\ell} = S^{3,1}_{j,k,\ell} = q(k, \ell, j),$$

$$S^{2,3}_{j,k,\ell} = S^{3,2}_{j,k,\ell} = q(j, k, \ell),$$

$$q(a,b,c) = \frac{\lambda + \mu}{4} \begin{cases} 1, & a = 2 \wedge (b = c = 1 \vee b = c = 3), \\ -1, & a = 2 \wedge (b = 1 \wedge c = 3 \vee b = 3 \wedge c = 1), \\ 0, & \text{else}. \end{cases}$$

Note that different boundary conditions result in different convolution schemes; see also Section 8.4. The non-zero pattern of the matrix A is shown in Fig. 9.6 for $n_1 = 5$, $n_2 = 7$, and $n_3 = 4$.

The remaining problem after discretization is to solve linear systems of equations $Au = f$. Note that the dimension of A can be huge. The registration of two n_1-by-n_2-by-n_3 images leads to $A \in \mathbb{R}^{(3N)\times(3N)}$, where $N = n_1 n_2 n_3$. For example, if the images are 256^3, we have to deal with approximately $5 \cdot 10^7$ unknowns. The registration of a part of the human brain presented at the end of this chapter results in linear systems with 52 428 800 unknowns. Thus, the usage and acceptance of this registration is strongly related to fast and efficient schemes for solving the linear system. In the next section we provide a solution scheme which is based on fast Fourier transformation techniques and leads to an $\mathcal{O}(N \log N)$ implementation.

9.5 Factorization of a structured matrix

Here, we identify a class of matrices which may be diagonalized by fast Fourier transformation (FFT) techniques. Our discussion follows that of Fischer & Modersitzki (1999). As it turns out, for dimension $d = 2, 3$, the matrices arising from the finite difference discretization of the Navier–Lamé equations introduced in the last section belong to this class if periodic boundary conditions are used.

Important ingredients for the following decompositions are the basic *circular matrix*

$$C_n := \begin{pmatrix} 0 & 1 & 0 & \cdots & 0 \\ \vdots & \ddots & \ddots & \ddots & \vdots \\ \vdots & \ddots & \ddots & \ddots & 0 \\ 0 & \ddots & \ddots & \ddots & 1 \\ 1 & 0 & \cdots & \cdots & 0 \end{pmatrix} \in \mathbb{C}^{n\times n}, \tag{9.16}$$

the nth *root of unity*

$$\omega_n := e^{-2\pi i/n}, \tag{9.17}$$

the Fourier matrix $F_n \in \mathbb{C}^{n\times n}$,

$$F_n := \frac{1}{\sqrt{n}} \left(\omega_n^{(j-1)(k-1)}\right)_{j,k=1,\dots,n}, \tag{9.18}$$

and

$$\Omega_n := \operatorname{diag}\left(\omega_n^0, \dots, \omega_n^{n-1}\right). \tag{9.19}$$

In this section we make intensive use of the Kronecker calculus; see, e.g., Horn & Johnson (1991, §4) or Brewer (1978).

9.5.1 *The three-dimensional case*

We start by defining this class of diagonalizable matrices for the three-dimensional case. The two-dimensional case is deduced later. Let $A \in \mathbb{R}^{(3n_1n_2n_3)\times(3n_1n_2n_3)}$ be a 3-by-3 block matrix with block circular matrices $A^{p,q}$,

$$A = \begin{pmatrix} A^{1,1} & A^{1,2} & A^{1,3} \\ A^{2,1} & A^{2,2} & A^{2,3} \\ A^{3,1} & A^{3,2} & A^{3,3} \end{pmatrix} \in \mathbb{R}^{(3n_1n_2n_3)\times(3n_1n_2n_3)}, \tag{9.20}$$

where for $p, q = 1, 2, 3$,

$$\begin{aligned} A^{p,q} &= C_{n_3}^{-1} \otimes A_3^{p,q} + I_{n_3} \otimes A_2^{p,q} + C_{n_3} \otimes A_1^{p,q} \\ &= \begin{pmatrix} A_2^{p,q} & A_1^{p,q} & & A_3^{p,q} \\ A_3^{p,q} & \ddots & \ddots & \\ & \ddots & A_2^{p,q} & A_1^{p,q} \\ A_1^{p,q} & & A_3^{p,q} & A_2^{p,q} \end{pmatrix} \in \mathbb{R}^{(n_1n_2n_3)\times(n_1n_2n_3)}. \end{aligned} \tag{9.21}$$

Each block $A_r^{p,q}$, $r = 1, 2, 3$, is block circular,

$$A_r^{p,q} = C_{n_2}^{-1} \otimes A_{r,3}^{p,q} + I_{n_2} \otimes A_{r,2}^{p,q} + C_{n_2} \otimes A_{r,1}^{p,q} \tag{9.22}$$
$$= \begin{pmatrix} A_{r,2}^{p,q} & A_{r,1}^{p,q} & & A_{r,3}^{p,q} \\ A_{r,3}^{p,q} & \ddots & \ddots & \\ & \ddots & A_{r,2}^{p,q} & A_{r,1}^{p,q} \\ A_{r,1}^{p,q} & & A_{r,3}^{p,q} & A_{r,2}^{p,q} \end{pmatrix} \in \mathbb{R}^{(n_1 n_2)\times(n_1 n_2)},$$

with circular blocks $A_{r,s}^{p,q}$, $s = 1, 2, 3$,

$$A_{r,s}^{p,q} = S_{3,s,r}^{p,q} C_{n_1}^{-1} + S_{2,s,r}^{p,q} I_{n_1} + S_{1,s,r}^{p,q} C_{n_1} \tag{9.23}$$
$$= \begin{pmatrix} S_{2,s,r}^{p,q} & S_{1,s,r}^{p,q} & & S_{3,s,r}^{p,q} \\ S_{3,s,r}^{p,q} & \ddots & \ddots & \\ & \ddots & S_{2,s,r}^{p,q} & S_{1,s,r}^{p,q} \\ S_{1,s,r}^{p,q} & & S_{3,s,r}^{p,q} & S_{2,s,r}^{p,q} \end{pmatrix} \in \mathbb{R}^{n_1 \times n_1}.$$

The 3-by-3-by-3 arrays $S^{p,q} = (S_{t,s,r}^{p,q})_{t,r,s=1,2,3}$, $p, q = 1, 2, 3$, are given and, in the case of elastic registration, coincide with the matrix stencils of the underlying discretization; cf., Table 9.1.

A main ingredient of our analysis is the fact that any circular matrix may be diagonalized by a Fourier matrix, a proof of which may be found for example in Davis (1979, Theorem 3.2.2).

Lemma 9.5 *Let $n \in \mathbb{N}$, and F_n and Ω_n be defined by eqns (9.18) and (9.19), respectively.*

1. *The Fourier matrix is unitary, $F_n^{-1} = F_n^{\mathrm{H}}$.*
2. *The circular matrix C_n (cf., eqn (9.16)) is diagonalized by F_n, $F_n^{\mathrm{H}} C_n F_n = \Omega_n$.*
3. *Any circular matrix $Z_n = \sum_{j=-n}^{n} \alpha_j C_n^j$ is diagonalized by F_n,*

$$F_n^{\mathrm{H}} Z_n F_n = \sum_{j=-n}^{n} \alpha_j (\Omega_n)^j.$$

Next, we show how to factorize the block matrix A; cf., eqn (9.20).

Lemma 9.6 *Let* $n_1, n_2, n_3 \in \mathbb{N}$, $p, q = 1, 2, 3$.

1. *The matrices* $A^{p,q}_{r,s} \in \mathbb{R}^{n_1 \times n_1}$ *(cf., eqn (9.23)) are diagonalized by* F_{n_1},

$$D^{p,q}_{r,s} := F^{\mathrm{H}}_{n_1} A^{p,q}_{r,s} F_{n_1},$$

$$D^{p,q}_{r,s} = \operatorname{diag}\big(D^{p,q}_{r,s;j},\ j = 1, \ldots, n_1\big),$$

$$\text{where}\quad D^{p,q}_{r,s;j} = S^{p,q}_{2,s,r} + S^{p,q}_{3,s,r}\overline{\omega}^{j-1}_{n_1} + S^{p,q}_{1,s,r}\omega^{j-1}_{n_1}.$$

2. *The matrices* $A^{p,q}_{r} \in \mathbb{R}^{(n_1 n_2) \times (n_1 n_2)}$ *(cf., eqn (9.22)) are diagonalized by* $F_{n_2} \otimes F_{n_1}$,

$$D^{p,q}_{r} := (F_{n_2} \otimes F_{n_1})^{\mathrm{H}} A^{p,q}_{r} (F_{n_2} \otimes F_{n_1}),$$

$$D^{p,q}_{r} = \operatorname{diag}\left(D^{p,q}_{r;k,j},\ \begin{matrix} j = 1, \ldots, n_1, \\ k = 1, \ldots, n_2 \end{matrix}\right),$$

$$\text{where}\quad D^{p,q}_{r;k,j} = D^{p,q}_{r,2;j} + \overline{\omega}^{k-1}_{n_2} D^{p,q}_{r,3;j} + \omega^{k-1}_{n_2} D^{p,q}_{r,1;j}.$$

3. *The matrices* $A^{p,q} \in \mathbb{R}^{(n_1 n_2 n_3) \times (n_1 n_2 n_3)}$ *(cf., eqn (9.21)) are diagonalized by* $F_{n_3} \otimes F_{n_2} \otimes F_{n_1}$,

$$D^{p,q} := (F_{n_3} \otimes F_{n_2} \otimes F_{n_1})^{\mathrm{H}} A^{p,q} (F_{n_3} \otimes F_{n_2} \otimes F_{n_1}),$$

$$D^{p,q} = \operatorname{diag}\left(D^{p,q}_{\ell,k,j},\ \begin{matrix} j = 1, \ldots, n_1, \\ k = 1, \ldots, n_2, \\ \ell = 1, \ldots, n_3 \end{matrix}\right),$$

$$\text{where}\quad D^{p,q}_{\ell,k,j} = D^{p,q}_{2;k,j} + \overline{\omega}^{\ell-1}_{n_3} D^{p,q}_{3;k,j} + \omega^{\ell-1}_{n_3} D^{p,q}_{1;k,j}.$$

4. *For the matrix* $A \in \mathbb{R}^{(3n_1 n_2 n_3) \times (3n_1 n_2 n_3)}$ *(cf., eqn (9.20)) we have*

$$F^{\mathrm{H}} A F = \begin{pmatrix} D^{1,1} & D^{1,2} & D^{1,3} \\ D^{2,1} & D^{2,2} & D^{2,3} \\ D^{3,1} & D^{3,2} & D^{3,3} \end{pmatrix}$$

where $F = I_3 \otimes F_{n_3} \otimes F_{n_2} \otimes F_{n_1}$.

Proof

1. Follows directly from Lemma 9.5.
2. From

$$
\begin{aligned}
B_r^{p,q} &:= (I_{n_2} \otimes F_{n_1})^{\mathrm{H}} A_r^{p,q} (I_{n_2} \otimes F_{n_1}) \\
&= C_{n_2}^{-1} \otimes \left(F_{n_1}^{\mathrm{H}} A_{r,3}^{p,q} F_{n_1}\right) \\
&\quad + I_{n_2} \otimes \left(F_{n_1}^{\mathrm{H}} A_{r,2}^{p,q} F_{n_1}\right) + C_{n_2} \otimes \left(F_{n_1}^{\mathrm{H}} A_{r,1}^{p,q} F_{n_1}\right) \\
&= C_{n_2}^{-1} \otimes D_{r,3}^{p,q} + I_{n_2} \otimes D_{r,2}^{p,q} + C_{n_2} \otimes D_{r,1}^{p,q}
\end{aligned}
$$

we deduce that $B_r^{p,q}$ is a block circular matrix with diagonal blocks. A suitable permutation P yields

$$
P^{-1} B_r^{p,q} P = \operatorname{diag}\left(L_{r,1}^{p,q}, \ldots, L_{r,n_1}^{p,q}\right),
$$

where $L_{r;j}^{p,q} = C_{n_2}^{-1} D_{r,3;j}^{p,q} + I_{n_2} D_{r,2;j}^{p,q} + C_{n_2} D_{r,1;j}^{p,q} \in \mathbb{R}^{n_2 \times n_2}$ and the permutation maps $(1, 2, \ldots, n_1,\ n_1+1, n_1+2, \ldots, n_1^2)$ to $(1, n_1+1, \ldots, n_1(n_1-1)+1,\ 2, n_1+2, \ldots, n_1^2)$. Finally, $L_{r;j}^{p,q}$ may be diagonalized by F_{n_2} showing that

$$
D_{r;k,j}^{p,q} = D_{r,2;j}^{p,q} + \overline{\omega}_{n_2}^{k-1} D_{r,3;j}^{p,q} + \omega_{n_2}^{k-1} D_{r,1;j}^{p,q},\ k = 1, \ldots, n_2,
$$

are the eigenvalues of $L_{r;j}^{p,q}$ and thus

$$
D_r^{p,q} = (F_{n_2} \otimes F_{n_1})^{\mathrm{H}} A_r^{p,q} (F_{n_2} \otimes F_{n_1}).
$$

3. From

$$
\begin{aligned}
B^{p,q} &:= (I_{n_3} \otimes F_{n_2} \otimes F_{n_1})^{\mathrm{H}} A^{p,q} (I_{n_3} \otimes F_{n_2} \otimes F_{n_1}) \\
&= C_{n_3}^{-1} \otimes D_3^{p,q} + I_{n_3} \otimes D_2^{p,q} + C_{n_3} \otimes D_1^{p,q}
\end{aligned}
$$

we deduce that also $B^{p,q}$ is a block circular matrix with diagonal blocks. A suitable permutation P yields

$$
P^{-1} B^{p,q} P = \operatorname{diag}\left(L_1^{p,q}, \ldots, L_{n_1 n_2}^{p,q}\right),
$$

where for $j = 1, \ldots, n_1$, $k = 1, \ldots, n_2$, and $m = (k-1)n_2 + j$,

$$
L_m^{p,q} = C_{n_3}^{-1} D_{3;k,j}^{p,q} + I_{n_3} D_{2;k,j}^{p,q} + C_{n_3} D_{1;k,j}^{p,q} \in \mathbb{R}^{n_3 \times n_3}.
$$

Finally, $L_m^{p,q}$ may be diagonalized by F_{n_3}.
4. The matrices $A^{p,q}$ are simultaneously diagonalizable. □

9.5.2 *The two-dimensional case*

For two-dimensional registration one is also interested in a 2-by-2 block matrix $A \in \mathbb{R}^{(2n_1n_2)\times(2n_1n_2)}$,

$$A = \begin{pmatrix} A_0^{1,1} & A_0^{1,2} \\ A_0^{2,1} & A_0^{2,2} \end{pmatrix} \in \mathbb{R}^{(2n_1n_2)\times(2n_1n_2)}, \tag{9.24}$$

where for $p, q = 1, 2$, the matrices $A_0^{p,q}$ are defined in eqn (9.22) using the 3-by-3 stencils $S^{p,q}$, where, for convenience, we set $S_{t,s}^{p,q} := S_{t,s,0}^{p,q}$, $t, s = 1, 2, 3$.

The eigendecomposition of the matrix for the two-dimensional registration can be deduced from Lemma 9.6 and is summarized in the following corollary.

Corollary 9.7 *For the matrix $A \in \mathbb{R}^{(2n_1n_2)\times(2n_1n_2)}$ (cf., eqn (9.24)) we have*

$$F^{\mathrm{H}}AF = \begin{pmatrix} D_0^{1,1} & D_0^{1,2} \\ D_0^{2,1} & D_0^{2,2} \end{pmatrix}$$

where $F = I_2 \otimes F_{n_2} \otimes F_{n_1}$, $A_0^{p,q}$ is given by eqn (9.22),

$$D_0^{p,q} := (F_{n_2} \otimes F_{n_1})^{\mathrm{H}} A_0^{p,q} (F_{n_2} \otimes F_{n_1}),$$

$$D_0^{p,q} = \operatorname{diag}\left(D_{0;k,j}^{p,q}, \begin{array}{l} j = 1, \ldots, n_1, \\ k = 1, \ldots, n_2 \end{array} \right),$$

where $$D_{0;k,j}^{p,q} = D_{0,2;j}^{p,q} + \bar{\omega}_{n_2}^{k-1} D_{0,3;j}^{p,q} + \omega_{n_2}^{k-1} D_{0,1;j}^{p,q},$$

and $$D_{0,s;j}^{p,q} = S_{2,s}^{p,q} + S_{3,s}^{p,q} \bar{\omega}_{n_1}^{j-1} + S_{1,s}^{p,q} \omega_{n_1}^{j-1}.$$

9.5.3 *Inversion of the structured matrices*

Our goal is to explicitly invert $F^{\mathrm{H}}AF$. The trouble is that the matrix A might be singular (and for the particular choices of S^p we are interested in, it is). Therefore we compute the so-called Moore–Penrose pseudo-inverse; see, e.g., Golub & van Loan (1989). To begin with, we investigate the Moore–Penrose pseudo-inverse $B^\dagger$ of a 2-by-2 matrix B.

Lemma 9.8 *For the symmetric matrix $B = \begin{pmatrix} b^1 & b^2 \\ b^2 & b^4 \end{pmatrix} \in \mathbb{R}^{2\times 2}$ the Moore–Penrose pseudo-inverse $B^\dagger$ is given by*

$$B^\dagger = \begin{cases} B^{-1}, & \text{for } \det(B) \neq 0, \\ 1/(b^1 + b^4)^2 \cdot B, & \text{for } \det(B) = 0,\ b^1 \neq 0 \vee b^4 \neq 0, \\ \mathbf{0}, & \text{otherwise.} \end{cases}$$

Proof

1. If $\det(B) \neq 0$, then $B^\dagger = B^{-1}$.
2. Let $\det(B) = 0$ but $b^1 \neq 0 \vee b^4 \neq 0$. Then B has a singular value decomposition $B = Q \operatorname{diag}(b^1 + b^4, 0) Q^\top$, where

$$Q = \frac{1}{\sqrt{(b^1)^2 + (b^2)^2}} \begin{pmatrix} b^1 & b^2 \\ b^2 & -b^1 \end{pmatrix} \quad \text{for } b^1 \neq 0,$$

$$Q = \frac{1}{\sqrt{(b^4)^2 + (b^2)^2}} \begin{pmatrix} b^2 & b^4 \\ b^4 & -b^2 \end{pmatrix} \quad \text{for } b^1 = 0.$$

   Hence, the pseudo-inverse is given by $B^\dagger = 1/(b^1 + b^4)^2 B$.
3. Let $\det(B) = 0$ and $b^1 = b^4 = 0$. Thus, $B = \mathbf{0}$ and $B^\dagger = \mathbf{0}$. □

We are now ready to present our main result. For the computation of the pseudo-inverse of a symmetric 3-by-3 matrix we forgo a lengthy formula and instead make use of numerical schemes.

Theorem 9.9

1. *For* $d = 2$, *let* A *be defined by eqn (9.24), where* $S^{p,q} \in \mathbb{R}^{3\times 3}$ *are such that* $D_0^{p,q}$ *are real and* $D_0^{21} = D_0^{12}$. *Then* $A^\dagger = F D^\dagger F^H$, *where*

$$D^\dagger := \begin{pmatrix} D_0^{1,1,\dagger} & D_0^{1,2,\dagger} \\ D_0^{1,2,\dagger} & D_0^{2,2,\dagger} \end{pmatrix}$$

   and the diagonal matrices $D_0^{p,q,\dagger} := \operatorname{diag}(D_{0;k,j}^{p,q,\dagger})$ *are defined by*

$$\begin{pmatrix} D_{0;k,j}^{1,1,\dagger} & D_{0;k,j}^{1,2,\dagger} \\ D_{0;k,j}^{1,2,\dagger} & D_{0;k,j}^{2,2,\dagger} \end{pmatrix} := \begin{pmatrix} D_{0;k,j}^{1,1} & D_{0;k,j}^{1,2} \\ D_{0;k,j}^{1,2} & D_{0;k,j}^{2,2} \end{pmatrix}^\dagger, \quad \begin{array}{l} j = 1, \ldots, n_1, \\ k = 1, \ldots, n_2. \end{array}$$

2. *For* $d = 3$, *let* A *be defined by eqn (9.20), where* $S^{p,q} \in \mathbb{R}^{3\times 3\times 3}$ *are such that* $D^{p,q}$ *are real, and* $D^{p,q} = D^{q,p}$ *for* $p, q = 1, 2, 3$. *Then* $A^\dagger = F D^\dagger F^{\mathrm{H}}$, *where*

$$D^\dagger := \begin{pmatrix} D^{1,1,\dagger} & D^{1,2,\dagger} & D^{1,3,\dagger} \\ D^{2,1,\dagger} & D^{2,2,\dagger} & D^{2,3,\dagger} \\ D^{3,1,\dagger} & D^{3,2,\dagger} & D^{3,3,\dagger} \end{pmatrix}$$

   and the diagonal matrices $D^{p,q,\dagger} = \operatorname{diag}(D_{\ell,k,j}^{p,q,\dagger})$ *are defined by*

$$\begin{pmatrix} D_{\ell,k,j}^{1,1,\dagger} & D_{\ell,k,j}^{1,2,\dagger} & D_{\ell,k,j}^{1,3,\dagger} \\ D_{\ell,k,j}^{2,1,\dagger} & D_{\ell,k,j}^{2,2,\dagger} & D_{\ell,k,j}^{2,3,\dagger} \\ D_{\ell,k,j}^{3,1,\dagger} & D_{\ell,k,j}^{3,2,\dagger} & D_{\ell,k,j}^{3,3,\dagger} \end{pmatrix} := \begin{pmatrix} D_{\ell,k,j}^{1,1} & D_{\ell,k,j}^{1,2} & D_{\ell,k,j}^{1,3} \\ D_{\ell,k,j}^{2,1} & D_{\ell,k,j}^{2,2} & D_{\ell,k,j}^{2,3} \\ D_{\ell,k,j}^{3,1} & D_{\ell,k,j}^{3,2} & D_{\ell,k,j}^{3,3} \end{pmatrix}^\dagger, \quad \begin{array}{l} j = 1, \ldots, n_1, \\ k = 1, \ldots, n_2, \\ \ell = 1, \ldots, n_3. \end{array}$$

Proof We only prove part 1. The proof of part 2 is along the same lines. The starting point is Lemma 9.6 and in particular Corollary 9.7. A suitable permutation matrix P yields

$$P^{-1}\begin{pmatrix} D_0^{1,1} & D_0^{1,2} \\ D_0^{1,2} & D_0^{2,2} \end{pmatrix} P = \operatorname{diag}\left(B_{k,j}, \begin{array}{c} j = 1, \ldots, n_1, \\ k = 1, \ldots, n_2 \end{array} \right),$$

where

$$B_{k,j} = \begin{pmatrix} D_{0;k,j}^{1,1} & D_{0;k,j}^{1,2} \\ D_{0;k,j}^{1,2} & D_{0;k,j}^{2,2} \end{pmatrix} \in \mathbb{R}^{2\times 2}.$$

The pseudo-inverses of $B_{k,j}$ are given by Lemma 9.8. □

9.5.4 *Inversion of the discrete Navier–Lamé equations*

In view of Theorem 9.9, the solution of the linear system in Algorithm 8.2 may be computed explicitly by applying a number of FFTs and by computing the pseudo-inverse of the block diagonal matrix D. The implementation details for this process are summarized in Algorithm 9.10 for dimension $d = 2$ and in Algorithm 9.11 for dimension $d = 3$.

We note that for the particular choices of the stencils S (cf., eqn (9.15) and Table 9.1), the computation of $D^\dagger$ simplifies considerably. For dimension $d = 2$, we have

$$\begin{aligned}
D_{0;k,j}^{1,1} &= S_{2,2}^{1,1} + 2\cos(2\pi(j-1)/n_1)S_{1,2}^{1,1} + 2\cos(2\pi(k-1)/n_2)S_{2,1}^{1,1}, \\
D_{0;k,j}^{2,2} &= S_{2,2}^{1,1} + 2\cos(2\pi(k-1)/n_2)S_{1,2}^{1,1} + 2\cos(2\pi(j-1)/n_1)S_{2,1}^{1,1}, \\
D_{0;k,j}^{1,2} &= D_{0;k,j}^{2,1} = -4S_{1,1}^{1,2}\sin(2\pi(j-1)/n_1)\sin(2\pi(k-1)/n_2),
\end{aligned}$$

and for dimension $d = 3$, we have

$$\begin{aligned}
D_{\ell,k,j}^{1,1} &= S_{2,2,2}^{1,1} + 2\cos(2\pi(j-1)/n_1)S_{1,2,2}^{1,1} \\
&\quad + 2S_{2,1,2}^{1,1}(\cos(2\pi(k-1)/n_2) + \cos(2\pi(\ell-1)/n_3)), \\
D_{\ell,k,j}^{2,2} &= S_{2,2,2}^{1,1} + 2\cos(2\pi(k-1)/n_2)S_{1,2,2}^{1,1} \\
&\quad + 2S_{2,1,2}^{1,1}(\cos(2\pi(j-1)/n_1) + \cos(2\pi(\ell-1)/n_3)), \\
D_{\ell,k,j}^{3,3} &= S_{2,2,2}^{1,1} + 2\cos(2\pi(\ell-1)/n_3)S_{1,2,2}^{1,1} \\
&\quad + 2S_{2,1,2}^{1,1}(\cos(2\pi(j-1)/n_1) + \cos(2\pi(k-1)/n_2)), \\
D_{\ell,k,j}^{1,2} &= D_{\ell,k,j}^{2,1} = -4S_{1,1,2}^{1,2}\sin(2\pi(j-1)/n_1)\sin(2\pi(k-1)/n_2), \\
D_{\ell,k,j}^{1,3} &= D_{\ell,k,j}^{3,1} = -4S_{1,1,2}^{1,2}\sin(2\pi(j-1)/n_1)\sin(2\pi(\ell-1)/n_3), \\
D_{\ell,k,j}^{2,3} &= D_{\ell,k,j}^{3,2} = -4S_{1,1,2}^{1,2}\sin(2\pi(k-1)/n_2)\sin(2\pi(\ell-1)/n_3).
\end{aligned}$$

Algorithm 9.10 *Algorithm for solving the discrete two-dimensional Navier–Lamé equations using a standard 2d FFT.*

Compute $\tilde{F}^1 := \texttt{fft2}(f_1(X))$ and $\tilde{F}^2 := \texttt{fft2}(f_2(X))$.
For $j = 1, \dots, n_1$, for $k = 1, \dots, n_2$,
$\quad \tilde{U}^1_{k,j} = D^{1,1,\dagger}_{k,j} \tilde{F}^1_{k,j} + D^{1,2,\dagger}_{k,j} \tilde{F}^2_{k,j}$,
$\quad \tilde{U}^2_{k,j} = D^{2,1,\dagger}_{k,j} \tilde{F}^1_{k,j} + D^{2,2,\dagger}_{k,j} \tilde{F}^2_{k,j}$,
end.
Compute $U_1 := \texttt{fft2}^{-1}(\tilde{U}^1)$, $U_2 := \texttt{fft2}^{-1}(\tilde{U}^2)$.

Algorithm 9.11 *Algorithm for solving the discrete three-dimensional Navier–Lamé equations using a standard 3d FFT.*

For $r = 1, 2, 3$,
$\quad$ compute $\tilde{F}^r := \texttt{fft3}(f_r(X))$,
end.
For $j = 1, \dots, n_1$, for $k = 1, \dots, n_2$, for $\ell = 1, \dots, n_3$,
$\quad$ solve

$$\begin{pmatrix} \tilde{U}^1_{\ell,k,j} \\ \tilde{U}^2_{\ell,k,j} \\ \tilde{U}^3_{\ell,k,j} \end{pmatrix} = \begin{pmatrix} D^{1,1}_{\ell,k,j} & D^{1,2}_{\ell,k,j} & D^{1,3}_{\ell,k,j} \\ D^{2,1}_{\ell,k,j} & D^{2,2}_{\ell,k,j} & D^{2,3}_{\ell,k,j} \\ D^{3,1}_{\ell,k,j} & D^{3,2}_{\ell,k,j} & D^{3,3}_{\ell,k,j} \end{pmatrix}^{\dagger} \begin{pmatrix} \tilde{F}^1_{\ell,k,j} \\ \tilde{F}^2_{\ell,k,j} \\ \tilde{F}^3_{\ell,k,j} \end{pmatrix},$$

end, end, end.
For $r = 1, 2, 3$,
$\quad$ compute $U^r := \texttt{fft3}^{-1}(\tilde{U}^r)$,
end.

Theorem 9.12 *Let $F_r \in \mathbb{R}^{n_1 \times \cdots \times n_d}$, $r = 1, \dots, d$, be given right hand sides and $N = n_1 \cdots n_d$. The numerical complexity of the outlined scheme for the solution of the discrete Navier–Lamé equations*

$$A(U_1, \dots, U_d)^\top = (F_1, \dots, F_d)^\top$$

with periodic boundary conditions is $\mathcal{O}(N \log N)$.

Proof In addition to the application of a d-dimensional FFT and its inverse, N small linear systems of size d-by-d have to be solved. □

9.6 Solving the Navier–Lamé equations

Using the fixed-point iteration (8.6), and the finite difference approximation of the Navier–Lamé operator with periodic boundary conditions as introduced in the previous section, we end up with the overall algorithm summarized in Algorithm 8.2.

For the computation of the force we use a bi- or tri-linear interpolation scheme; cf., Section 3.1.3. The time consuming part is, however, the solution of the linear system, which has to be done in each iteration. Exploiting the direct solver introduced in this chapter gives an $\mathcal{O}(N \log N)$ implementation. It is worthwhile noticing that iterative schemes may provide an attractive alternative to this solution process. However, using iterative solvers one also introduces an additional inner iteration and computes only an approximation to the solution of the linear system.

9.7 Regularizing the updates of the displacements

A very brief summary of elastic registration reads as follows. Given $u^{(k)}$, compute the force $f(\,\cdot\,, u^{(k)}(\cdot))$, and obtain the new update $u^{(k+1)}$ from $u^{(k+1)} = \mathcal{A}^{\dagger} f(\,\cdot\,, u^{(k)}(\,\cdot\,))$. The final deformation is given by $x + u^{(K)}(x)$, where K denotes the number of iterations needed to fulfill a certain stopping criterion.

Though this approach presents a physically meaningful model of an elastic tissue, it appears to be too restrictive for many image registration problems. The remedy is to use a regularized incremental update. Here, the final deformation is given by $x + \sum_{k=1}^{K} u^{(k)}(x)$, where $u^{(k+1)} = \mathcal{A}^{\dagger} f(\,\cdot\,, u^{(k)}(\,\cdot\,))$. Using this idea, the overall elastic deformation is decomposed into K small steps, where the elastic memory of the body is erased from step to step. This variant is used for the examples shown in the following section.

A more general approach which is related to optical flow computation is to introduce a flow $I(x,t)$ and to minimize

$$\mathcal{J}[u] = \int_0^1 \int_\Omega \left(\frac{d}{dt} I(x - u(x,t), t) \right)^2 dx \; + \alpha \mathcal{S}[u] \, dt$$

where $I(x,0) = R(x)$, $I(x,1) = T(x)$, and additional boundary conditions for $I(\partial\Omega, t)$ are to be imposed; cf., Keeling & Ring (2002).

9.8 Elastic registration: • to C

We start with an academic example, given the two images depicted in Fig. 9.7. A MATLAB description of both m-by-m images is given in Table 9.2. A warping

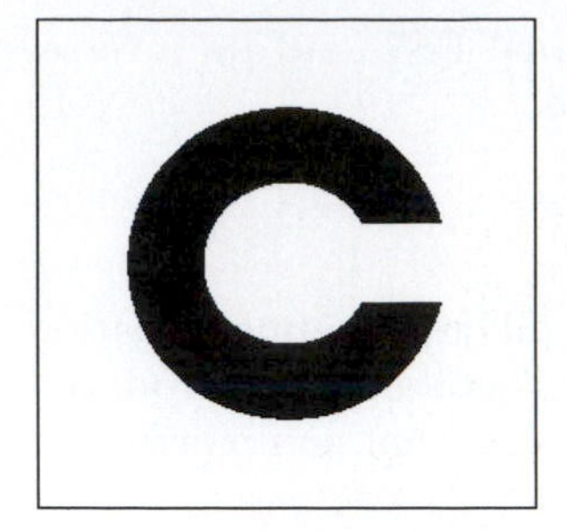

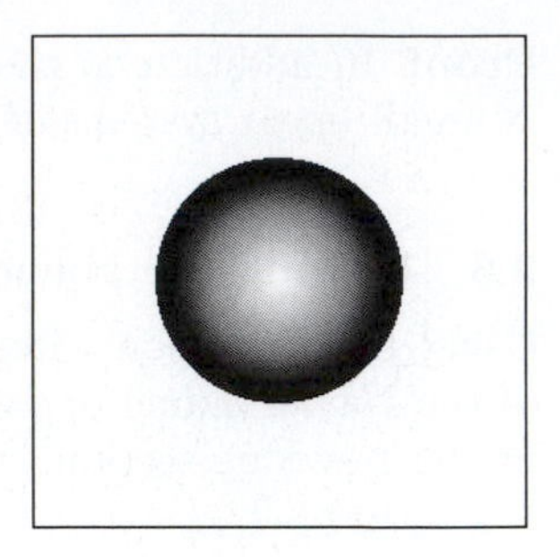

FIG. 9.7 Reference C, template •, and shaded template.

Table 9.2 MATLAB *code for reference* C *and template* •.

```
[x,y] = meshgrid(linspace(-1,1,m));
R = zeros(m);
R(x.^2+y.^2>0.33^2 & x.^2+y.^2<0.64^2) = 255;
R(x>0 & abs(y)<0.16) = 0;
T = 0.5^2 - x.^2 - y.^2;
T(T>0) = 255;
```

from • to C is impossible with an elastic deformation, because the images differ by too much.

In Fig. 9.8 we visualize the deformations for the elastic registrations with Dirichlet, Neumann, and periodic boundary conditions. Obviously, the boundary conditions have an impact on the deformations particularly near the boundary but also globally. However, since we assume the images to be embedded completely in a background, this differences are regarded as irrelevant for the overall registration goal.

In Fig. 9.9, we show intermediate results for $k = 0(20)100$. In order to visualize the deformation, the template has been shaded. Note that the computations have been performed on the original images. In this example, the Lamé constants are $\lambda = 0$ and $\mu = 5000$. For this example we have numerical convergence of the scheme after about 100 iterations, cf., Fig. 9.8 for periodic boundary conditions.

9.9 Elastic registration: hands

The next example is slightly more realistic. Figure 9.10 shows two modified X-ray images of human hands. We also show the linearly registered template, the linearly and elastically registered template, and the differences of the original, linearly and elastically deformed template with the reference image. For this example, we chose the Lamé constants $\mu = 500$ and $\lambda = 0$. As is apparent from this figure, the additional elastic registration reduces the image difference considerably. Intermediate registration results are depicted in Fig. 9.11.

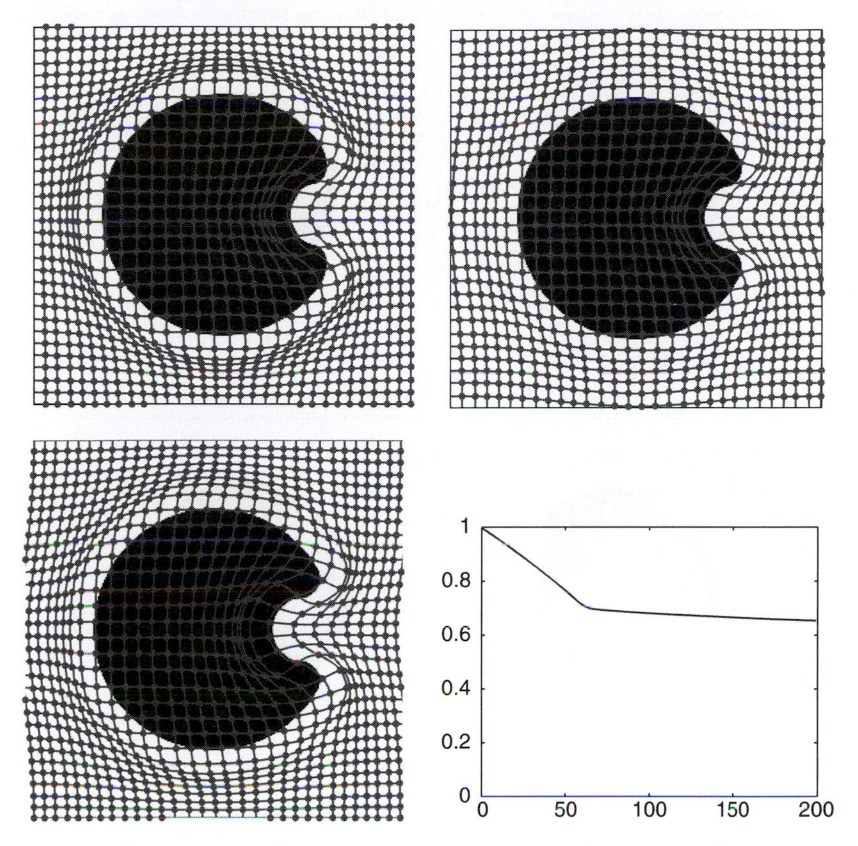

FIG. 9.8 Deformations after elastic registration: (TOP LEFT) with Dirichlet boundary conditions (TOP RIGHT) with Neumann boundary conditions (BOTTOM LEFT) with periodic boundary conditions (BOTTOM RIGHT) convergence history for periodic boundary conditions.

9.10 Elastic registration: brain

An example of a two-dimensional elastic registration has already been presented in Chapter 2, see Fig. 2.4. We now also present a registration of a part of the human brain covering the visual cortex, see Fig. 9.12.

After an appropriate pre-registration as explained in Part I, each section $S^{(\nu)}$, $\nu = 1, \ldots, 100$, has been registered with $\varphi^{(\nu)}$, where the $\varphi^{(\nu)}$'s are such that

$$\sum_{\nu=2}^{100} \left\| S^{(\nu)} \circ \varphi^{(\nu)} - S^{(\nu-1)} \circ \varphi^{(\nu-1)} \right\|_{L_2(\Omega)}^2 + \sum_{\nu=1}^{100} \mathcal{S}^{\text{elas}}[\varphi^{(\nu)}] \longrightarrow \min .$$

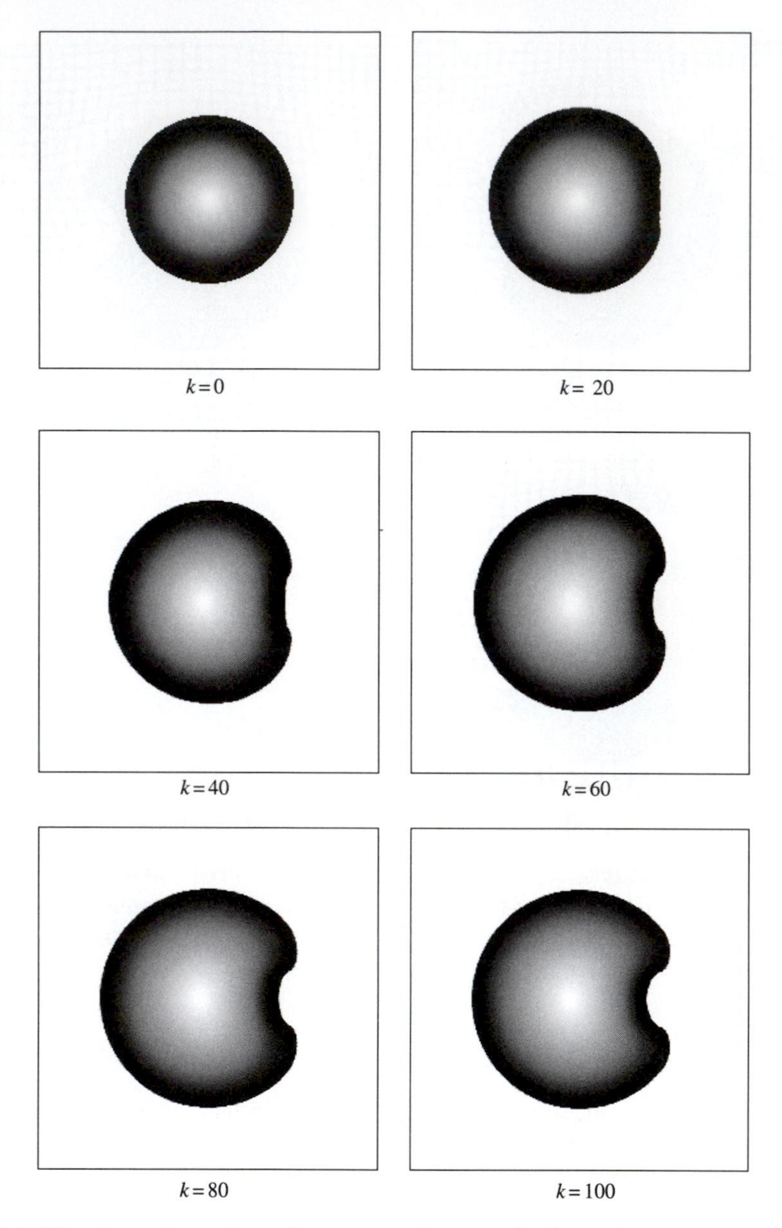

FIG. 9.9 Elastic registration of • to C with periodic boundary conditions, $\mu = 5000$, and $\lambda = 0$; intermediate results for $k = 0(20)100$.

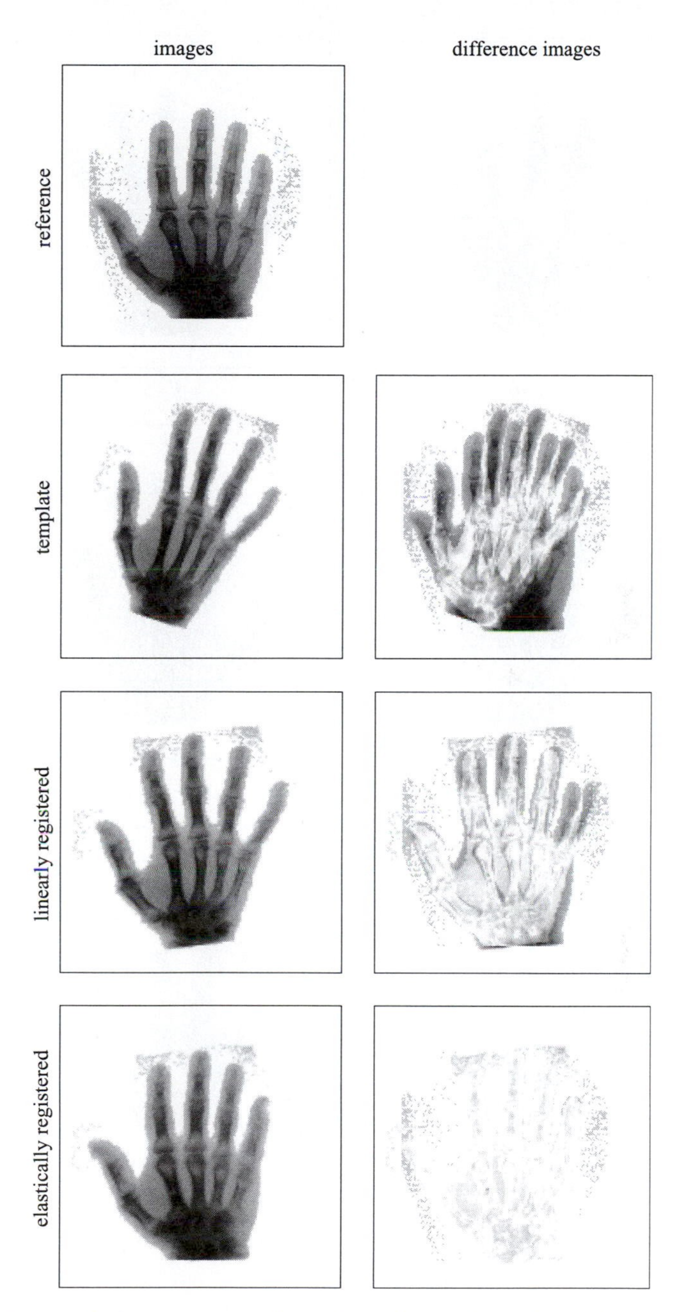

FIG. 9.10 LEFT: Reference, template, linearly registered template, and elastically registered template; RIGHT: difference images.

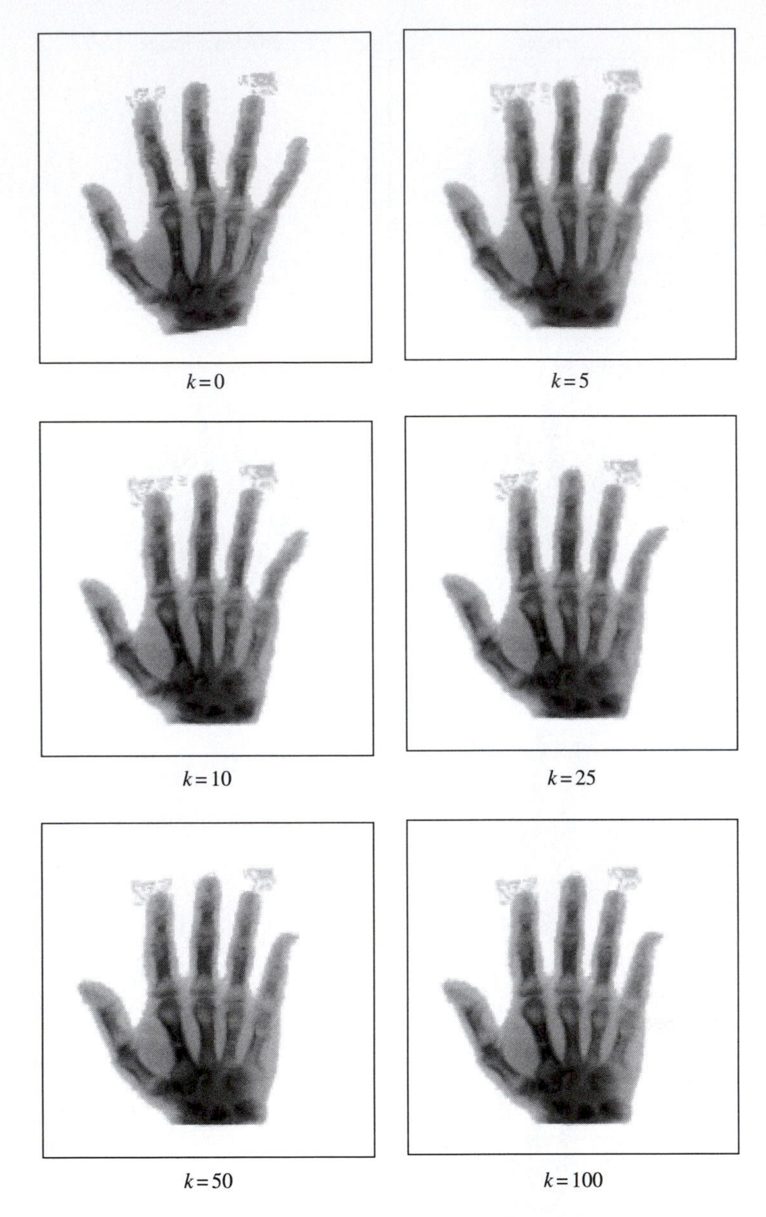

FIG. 9.11 Elastic registration of hands, $\mu = 500$, $\lambda = 0$; intermediate results for various values of k.

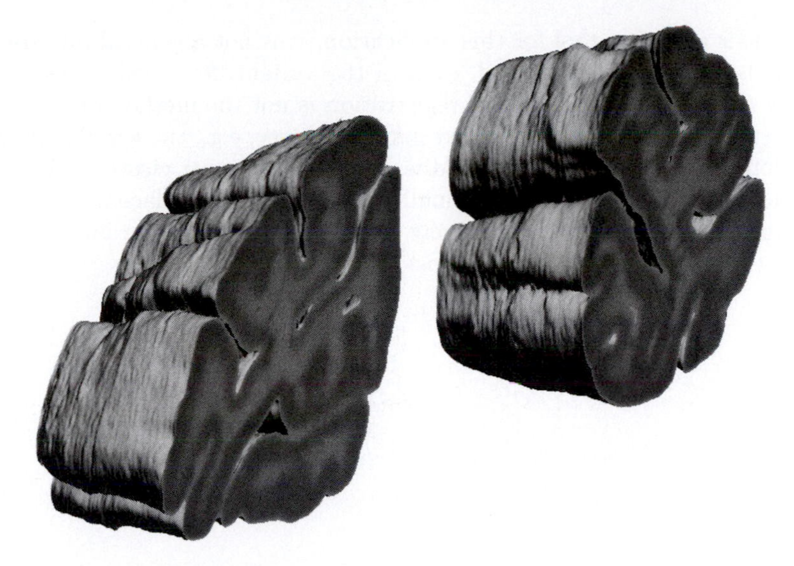

FIG. 9.12 Elastic registration of a part of a histological serial sectioning of a human brain; 100 registered sections in a 512^2 resolution.

After a discretization as described in the previous sections, this leads to a system of non-linear equation for the discrete $\vec{U}^{(\nu)}$,

$$[I_{100} \otimes A](\vec{U}^{(1)}, \dots, \vec{U}^{(100)})^\top = (\vec{F}^{(1)}, \dots, \vec{F}^{(100)})^\top,$$

where A is defined by eqn (9.24). Here we used a non-linear Gauss–Seidel iteration, i.e., roughly speaking we define

$$F^{(\nu,k)} = (S^{(\nu-1)} \circ \varphi^{(\nu-1,k+1)} - 2S^{(\nu)} \circ \varphi^{(\nu,k)} + S^{(\nu+1)} \circ \varphi^{(\nu+1,k)}) \nabla S^{(\nu)} \circ \varphi^{(\nu,k)}$$

and iterate $AU^{(\nu,k+1)} = F^{(\nu,k)}$ with respect to k. A multi-resolution of the images based on a Gaussian pyramid is used. For the solution of the linear systems we apply the direct solution technique developed in this chapter.

As is apparent from Fig. 9.12 this registration results in a smooth and anatomically meaningful reconstruction of this region of the brain. For a discussion of the anatomical issues, we refer to Schmitt (2001).

9.11 Discussion of elastic registration

Elastic registration presents an interesting registration tool. The driving force of the registration is the minimization of a distance measure and the transformation is restricted to the class of elastic deformations.

This technique seems to be appropriate for the registration of images arising from a sectioning of an elastic material, e.g., within the HNSP (cf., Chapter 2).

Though it is a perfect tool for this application, it is not a general purpose tool. The regularizer is very restricted, even in the variant described in Section 9.7. For very dissimilar objects, elastic registration is not the method of choice. The distance measure may not be reduced sufficiently; see, e.g., the • to C registration in Section 9.8. Note that the method is designed for linear elastic deformations only and implicitly assumes only small changes of the displacement. Relaxing the Lamé constants by too much may violate this assumption and one might end up with a non-diffeomorphic transformation.

10

FLUID REGISTRATION

10.1 Variational motivation of fluid registration

The essential difference between elastic and fluid registration is, roughly speaking, that for elastic registration the regularization is based on $\mathcal{S}^{\text{elas}}[u] = \mathcal{P}[u]$, whereas in the fluid approach it is based on $\mathcal{S}^{\text{fluid}}[u] = \mathcal{P}[\partial_t u]$. Thus, if u converges to a steady-state solution, the regularizer becomes less important. In this context, the fluid approach may also be viewed as a sequence of Tychonov regularizations for the distance measure,

$$\mathcal{D}[u] + \alpha^k \mathcal{S}[u] \xrightarrow{u} \min, \quad k \geq 0,$$

where α^k goes to zero.

From Theorem 9.4 we obtain

$$f(\,\cdot\,, u(\,\cdot\,,\,\cdot\,)) = \mu \Delta v + (\lambda + \mu) \nabla \operatorname{div} v, \quad v = \partial_t u,$$

as a characterizing equation for fluid registration.

In contrast to the elastic model the strain in a fluid model depends on the rate of change. Elastic models are characterized by a spatial smoothing of the displacement field. In contrast, fluid models are characterized by a spatial smoothing of the velocity field. Thus, in principle, any displacement can be obtained given enough time: the internal stresses disappear eventually.

10.2 Physical motivation of fluid registration

In this section we present a coherent description of the elastic and fluid continuum models of the deformation of a body, see also Christensen (1994) or Bro-Nielsen (1996). Note that the fluid model is also called *visco-elastic*; cf., e.g., Wollny & Kruggel (2000).

The transformation $\varphi : \mathbb{R}^d \times \mathbb{R} \to \mathbb{R}$ (cf., Section 9.1.1) is now time dependent and smooth. For any time t, $\varphi(\cdot, t)$ has to be bijective and to preserve the orientation of the tissue. With the displacement $u : \mathbb{R}^d \times \mathbb{R} \to \mathbb{R}^d$ (cf., eqn (8.1)) the location of a particle P can be described by

$$\tilde{x} = \varphi(x, t) = x + u(x, t),$$

where $x := \varphi(P, 0)$ is the initial position of the particle and $\tilde{x}$ its current position at a fixed time t. With $\varphi(\cdot, t)$ being bijective, we can make use of its inverse

$\tilde{\varphi}(\cdot,t) := \varphi^{-1}(\cdot,t)$ and obtain

$$x = \tilde{\varphi}(\tilde{x},t) = \tilde{x} - \tilde{u}(\tilde{x},t),$$

where $u(\cdot,t)$ and $\tilde{u}(\cdot,t)$ are related via $u(x,t) = \tilde{u}(\tilde{x},t)$.

Using the initial coordinates x of the particles P as the reference coordinate system is also called the Lagrange frame, whereas tracking the particles with respect to their actual position $\tilde{x}$ is called the Euler frame; see also Section 3.3.2 and Remark 9.2. Since the displacement and the velocity of a particle are physical properties and thus independent of the reference frame, we have

$$u(x,t) = \tilde{u}(\tilde{x},t) \quad \text{and} \quad v(x,t) = \tilde{v}(\tilde{x},t),$$

where the velocity of the particle P is defined as the partial time derivative of the transformation,

$$v(x,t) := \partial_t \varphi(x,t), \quad x = \varphi(P,0).$$

Note that

$$\tilde{v}(\tilde{x},t) = \frac{d}{dt}\tilde{u}(\tilde{x},t) = \nabla\tilde{u}(\tilde{x},t)\partial_t\tilde{x} + \partial_t\tilde{u}(\tilde{x},t) = \nabla\tilde{u}(\tilde{x},t)\tilde{v}(\tilde{x},t) + \partial_t\tilde{u}(\tilde{x},t).$$

As already introduced in Chapter 9, *strain* is measured using the deformation gradient $\nabla\varphi = I_d + \nabla u$; cf., Section 9.1.2.

10.2.1 *The basics of fluid mechanics*

The continuous model satisfies the basic physical axioms such as balance of mass and linear and angular momentum, which are stated below. An important relation between Lagrange and Euler coordinates is given by the Reynolds' transport theorem; cf., e.g., Gurtin (1981, III.10 (5)).

Theorem 10.1 (Reynolds' transport theorem) *Let $\varphi : \mathbb{R}^d \times \mathbb{R} \to \mathbb{R}^d$ be a smooth spatial field. Then for any set Ω and time t,*

$$\begin{aligned} &\frac{d}{dt}\int_{\varphi(\Omega,t)} g(\tilde{x},t)\,d\tilde{x} \\ &\quad = \int_{\varphi(\Omega,t)} \frac{d}{dt}g(\tilde{x},t) + g(\tilde{x},t)\operatorname{div}\tilde{v}(\tilde{x},t)\,d\tilde{x} \\ &\quad = \int_{\varphi(\Omega,t)} \partial_t g(\tilde{x},t) + \operatorname{div}[g(\tilde{x},t)\tilde{v}(\tilde{x},t)]\,d\tilde{x}. \end{aligned}$$

Conservation of mass Let φ denote the density of the body Ω and let $\tilde{\rho}$ be related to ρ via $\tilde{\rho}(\tilde{x},t) := \rho(x,t)$. The total mass of the body is constant, i.e.,

$$\frac{d}{dt}\int_{\varphi(K,t)} \tilde{\rho}(\tilde{x},t)\,d\tilde{x} = 0,$$

for any $K \subset \Omega$. Applying the transport theorem 10.1 and using the fact that the set K can be chosen arbitrarily, we obtain the *continuity equation*

$$\partial_t \tilde{\rho} + \operatorname{div}(\tilde{\rho}\tilde{v}) = 0. \tag{10.1}$$

If in particular the body is an incompressible continuum, the density $\tilde{\rho}$ is constant and the continuity equation (10.1) results in $\operatorname{div}\tilde{v} = 0$.

Conservation of linear momentum The axiom of conservation of linear momentum balances the momentum and the external and internal forces of any subbody K,

$$\frac{d}{dt}\int_{\varphi(K,t)} \tilde{\rho}\tilde{v}\,d\tilde{x} = \int_{\varphi(K,t)} \tilde{f}\,d\tilde{x} - \int_{\partial[\varphi(K,t)]} \tilde{\gamma}(\tilde{x},\vec{n})\,d\tilde{x},$$

where γ denotes the *stress* (cf., Section 9.1.3), $\partial[\varphi(K,t)]$ denotes the boundary of the set $\varphi(K,t)$, and $\vec{n}$ denotes the outer normal. Here, we use the Euler coordinates.

From the transport theorem 10.1 and the continuity equation (10.1) we have

$$\frac{d}{dt}\int_{\varphi(K,t)} \tilde{\rho}\tilde{v}\,d\tilde{x} = \int_{\varphi(K,t)} \tilde{\rho}\frac{d}{dt}\tilde{v} + \tilde{v}\left(\frac{d}{dt}\tilde{\rho} + \tilde{\rho}\operatorname{div}\tilde{v}\right)d\tilde{x} = \int_{\varphi(K,t)} \tilde{\rho}\frac{d}{dt}\tilde{v}\,d\tilde{x}.$$

With the stress tensor Σ (cf., Section 9.1.3), we thus have

$$\int_{\varphi(K,t)} \left(\tilde{\rho}\frac{d}{dt}\tilde{v} - \tilde{f} + \operatorname{div}\tilde{\Sigma}\right)d\tilde{x} = 0,$$

or, varying over all subbodies,

$$\tilde{\rho}\frac{d}{dt}\tilde{v} = \tilde{f} - \operatorname{div}\tilde{\Sigma}. \tag{10.2}$$

Remark 10.1 For *static elastic deformations* all time derivatives are zero, since there is no temporal change. Thus, if the strain–stress relation discussed in the previous chapter is used, eqn (10.2) becomes

$$\tilde{f} = \operatorname{div}\tilde{\Sigma} = \mu\Delta\tilde{u} + (\lambda+\mu)\nabla\operatorname{div}\tilde{u}$$

(see Section 9.1.4) and we rediscover the Navier–Lamé equations.

10.2.2 *Stokes fluids*

The Stokes hypothesis is that the stress tensor can be decomposed into parts related to velocity $\tilde{v}$ and the pressure $\tilde{p}$,

$$\tilde{\Sigma} = [\mu(\nabla\tilde{v} + \nabla\tilde{v}^\top) + \lambda \operatorname{div} \tilde{v} I_d] + \tilde{p} I_d. \tag{10.3}$$

Assuming $\tilde{\Sigma}$ to be as in eqn (10.3) it follows that

$$\operatorname{div} \tilde{\Sigma} = \mu\Delta\tilde{v} + (\lambda + \mu)\nabla \operatorname{div} \tilde{v} + \nabla\tilde{p}.$$

Moreover, if we assume very slow motion flows (small Reynolds' number), we may neglect the time derivative of the velocity and the spatial derivative of the pressure. Hence, from eqn (10.2) we get

$$\tilde{f} = \operatorname{div} \tilde{\Sigma} = \mu\Delta\tilde{v} + (\lambda + \mu)\nabla \operatorname{div} \tilde{v},$$

which is the underlying partial differential equation for *fluid registration*. Note that severe simplifications of the general fluid model have been applied. In particular, one assumes that the momentum vanishes and the fluid fulfills the Stokes hypothesis with spatially constant pressure.

10.3 Solving the Navier–Lamé equations for the velocity

Now we present a numerical scheme for the solution of the partial differential equation

$$\mu\Delta v + (\lambda + \mu)\nabla \operatorname{div} v = f(\,\cdot\,, u(\,\cdot\,,\,\cdot\,)), \quad v = \partial_t u + \nabla u\; v; \tag{10.4}$$

see Section 10.2. Since we are only interested in the Euler coordinates, from now on we omit the additional $\tilde{}$-notation of the variables.

The basic idea here is to use a fixed-point iteration, i.e., to compute the force field f for a given displacement u, to solve the linear partial differential equation for a new velocity v, and to compute the new displacement from this velocity using an Euler scheme. For the numerical computations we use a finite difference approximation for the derivatives as introduced in Chapter 8. For each step k, the computation of the force field $\vec{F}^{(k)}$ and the solution of the linear system of equations $A\vec{V}^{(k+1)} = \vec{F}^{(k)}$ are the same as in our implementation of elastic registration. The only difference with respect to the implementation of elastic registration is the computation of $\vec{U}^{(k+1)}$ from $\vec{V}^{(k+1)}$, which requires an additional Euler step.

For each grid point $x_i \in \mathbb{R}^d$ with $i = (i_1, \ldots, i_d) \in \mathbb{N}^d$ we also have $u^{(k)}(x_i) \in \mathbb{R}^d$. Thus, we need an $n_1 \times \cdots \times n_d \times d$ array for the storage of all components of $u^{(k)}(X)$. This array is denoted by

$$\vec{U}^{(k)} = [U^{(k,1)}, \ldots, U^{(k,d)}] \in \mathbb{R}^{n_1 \times \cdots \times n_d \times d}.$$

For convenience we introduce this data structure also for the identity on the grid and denote the corresponding array by $\vec{X} = [X^1, \ldots, X^d]$. Hence, the coordinates of the ith grid point are $x_i = (X_i^1, \ldots, X_i^d)^\top$.

Example 10.1 Suppose $d = 2$, $n_1 = n_2 = 3$, and our grid matrix is given, for example, by

$$X = (x_{i_1,i_2})_{i_1,i_2=1,2,3} \in \mathbb{R}^{n_1 \times n_2} \quad \text{with} \quad x_{i_1,i_2} = (i_1, i_2)^\top \in \mathbb{R}^d.$$

With

$$X^1 = \begin{pmatrix} 1 & 1 & 1 \\ 2 & 2 & 2 \\ 3 & 3 & 3 \end{pmatrix}, \quad X^2 = \begin{pmatrix} 1 & 2 & 3 \\ 1 & 2 & 3 \\ 1 & 2 & 3 \end{pmatrix} \in \mathbb{R}^{3\times 3}$$

we have $\vec{X} = [X^1, X^2]$ and, for example, $x_{2,3} = [X_{2,3}^1, X_{2,3}^2]^\top = (2,3)^\top$.

For a fixed index $i = (i_1, \ldots, i_d)$, we set

$$\vec{U}_i^{(k)} := \left(U_i^{(k,1)}, \ldots, U_i^{(k,d)}\right)^\top \in \mathbb{R}^d$$

and

$$\vec{V}_i^{(k)} := \left(V_i^{(k,1)}, \ldots, V_i^{(k,d)}\right)^\top \in \mathbb{R}^d,$$

respectively. Using a centered finite difference approximation for $\nabla u(x_i, k\tau)$,

$$\hat{J}_i^{(k)} := \left(\hat{J}_{i;\ell,q}^{(k)}\right)_{\ell,q\,=\,1,\ldots,d}, \quad \text{where } \hat{J}_{i;\ell,q}^{(k)} := \tfrac{1}{2}\left(U_{i+e_q}^{(k,\ell)} - U_{i-e_q}^{(k,\ell)}\right), \tag{10.5}$$

and a forward finite difference approximation for the partial time derivative $\partial_t u(x_i, k\tau)$, we obtain a discretized version of the Euler step,

$$\frac{\vec{U}_i^{(k+1)} - \vec{U}_i^{(k)}}{\tau} = \left(I_d - \hat{J}_i^{(k)}\right)\vec{V}_i^{(k)} \quad \text{for all } i.$$

Our algorithm is summarized in Algorithm 10.2.

Example 10.2 Suppose $d = 2$, $n_1 = n_2 = 3$, $\vec{U} = [U^{(k,1)}, U^{(k,2)}]$. For a fixed k and all $i = (i_1, i_2)$ we have to compute the d^2 values

$$\hat{J}_{i;1,1}^{(k)} = \tfrac{1}{2}\left(U_{i_1+1,i_2}^{(k,1)} - U_{i_1-1,i_2}^{(k,1)}\right), \quad \hat{J}_{i;1,2}^{(k)} = \tfrac{1}{2}\left(U_{i_1,i_2+1}^{(k,1)} - U_{i_1,i_2-1}^{(k,1)}\right),$$
$$\hat{J}_{i;2,1}^{(k)} = \tfrac{1}{2}\left(U_{i_1+1,i_2}^{(k,2)} - U_{i_1-1,i_2}^{(k,2)}\right), \quad \hat{J}_{i;2,2}^{(k)} = \tfrac{1}{2}\left(U_{i_1,i_2+1}^{(k,2)} - U_{i_1,i_2-1}^{(k,2)}\right).$$

The matrix

$$\hat{J}_i^{(k)} = \begin{pmatrix} \hat{J}_{i;1,1}^{(k)} & \hat{J}_{i;1,2}^{(k)} \\ \hat{J}_{i;2,1}^{(k)} & \hat{J}_{i;2,2}^{(k)} \end{pmatrix} \in \mathbb{R}^{d\times d}$$

is the finite difference approximation to $\nabla u(x_i, k\tau)$.

Algorithm 10.2 *MATLAB pseudo-code for fluid registration.*

$[\vec{U}, T_{\text{out}}] = \textbf{FluidRegistration}(R, T);$
  $k = 0;\quad q = \text{regrid\_counter} = 0;\quad [n_1, \ldots, n_d] = \text{size}(R);$
  $\vec{X} = [X_1, \ldots, X_d] = \text{ndgrid}(1 : n_1, \ldots, 1 : n_d);$
  $\vec{U}^{(k)} = \text{zeros}(\text{size}(\vec{X}));\quad \vec{Y}^{(q)} = \text{zeros}(\text{size}(\vec{X}));$
  setup_NLE_matrix;

  while 1
    $\vec{W}^{(k)} = \text{interpolate}(\vec{Y}^{(\text{regrid\_counter})}, \vec{X} - \vec{U}^{(k)});$
    $T^{(k)} = \text{interpolate}(T, \vec{X} - \vec{W}^{(k)} - \vec{U}^{(k)});$
    if $k \geq \text{maxiter}$, STOP; end;
    if $||R - T^{(k-1)}||_F - ||R - T^{(k)}||_F \leq ||R - T^{(k)}||_F \ \text{tol}_{\text{dist}}$, STOP; end;

    $\vec{F}^{(k)} = \text{compute\_forces}(R, T^{(k)});$
    if $||\vec{F}^{(k)}||_V \leq \text{tol}_{\text{for}}$, STOP; end;
    $\vec{V}^{(k)} = \text{solve\_NLE\_matrix}(\vec{F}^{(k)});$
    for $j, \ell = 1, \ldots, d,\quad J_{j,\ell}^{(k)} = \text{compute\_Jacobian}(\vec{U}_j^{(k)}, \ell);$ end,
    $\text{min\_Jacobian} = \min(\det([J_{j,\ell}^{(k)}]));$
    if ($\text{min\_Jacobian} < \text{tol}_{\text{Jac}}$),
      $\text{regrid\_counter} = \text{regrid\_counter} + 1;$
      $\vec{Y}^{(\text{regrid\_counter})} = \vec{W}^{(k)} + \vec{U}^{(k)};$
      $\vec{U}^{(k+1)} = \text{zeros}(\text{size}(\vec{X}));$
    else
      $\delta\vec{U}^{(k)} = [J_{j,\ell}^{(k)}]\vec{V}^{(k)};$
      $\delta u_{\max}^{(k)} = ||\delta\vec{U}^{(k)}||_V;$
      $\delta t^{(k)} = \min\{1, \text{tol}_u / \delta u_{\max}^{(k)}\};$
      $\vec{U}^{(k+1)} = \vec{U}^{(k)} + \delta t^{(k)} \delta\vec{U}^{(k)};$
    end,
    $k = k + 1;$
  end;
  $\vec{U} = \vec{W}^{(k)} + \vec{U}^{(k)};$
  $T_{\text{out}} = \text{interpolate}(T, \vec{X} - \vec{U}).$

Remark 10.2 It is worthwhile noticing that the Jacobian matrix

$$J(x) := \nabla[x - u(x)]$$

arising in the Euler step

$$\partial_t u = (\mathcal{I} - \nabla u)v \tag{10.6}$$

is nothing but the derivative of the transformation with respect to the Euler coordinates. Since the entries of this matrix are available during the computation, it is to easy to verify whether this matrix is regular or not. If the matrix is not regular, the transformation is not bijective and the step is not admissible. To circumvent this undesirable situation, Christensen (1994) suggests a so-called *regridding*, which means to clear the memory of the deformation.

Christensen suggests a regridding if $\det J(x) < \text{tol}$ for a grid point x. A mathematically preferable criterion is based on the smallest singular value of $J(x)$. Although of the same complexity as the computation of the determinants, the computation of the singular values is slightly more involved.

10.4 Implementation details

A few remarks concerning the implementation of our fluid registration scheme are in order. To this end we provide a MATLAB pseudo-code of our implementation, see Algorithm 10.2.

In an initialization phase, we set the iteration counter $k = 0$ and a counter for the regridding steps to zero. The d-dimensional images to be registered are $R, T \in \mathbb{R}^{n_1 \times \cdots \times n_d}$.

The initial displacement as well as the variable for the regridding step is set to zero, $\vec{U}^{(0)} = \vec{Y}^{(0)} = 0$. Finally, we initialize the matrices $D^{j,k}$, $j, k = 1, \ldots, d$; cf., Lemma 9.6.

In the following `while`-loop we find four essential blocks: the computation of the deformed template $T^{(k)}$, the computation of the forces $\vec{F}^{(k)}$, the computation of the velocities $\vec{V}$, and the Euler step. The most striking part is probably the regridding step.

10.4.1 *Euler step and regridding*

The displacement and the velocity are connected via the material derivative, see Section 10.3 and in particular eqn (10.6).

An approximation to the Jacobian of the transformation at the grid point x_i is given by

$$J_i^{(k)} = I_d - \hat{J}_i^{(k)}$$

where $\hat{J}_i^{(k)}$ is an approximation to the Jacobian $\nabla u(x_i, k\tau)$; cf., eqn (10.5).

The transformation is regarded as admissible if the determinants of the Jacobian on the grid exceed a certain user-supplied threshold tol_{Jac}. In our implementation we used $\text{tol}_{\text{Jac}} = 0.025$.

If the transformation is admissible, the Euler step $\vec{U}_i^{(k+1)} = \vec{U}_i^{(k)} + \delta t^{(k)} \delta \vec{U}_i^{(k)}$ is performed for all i, with

$$\delta \vec{U}_i^{(k)} = \left(I_d - \hat{J}_i^{(k)}\right) \vec{V}_i^{(k)}.$$

The time-step $\delta t^{(k)}$ is chosen such that $||\delta t^{(k)} \delta \vec{U}^{(k)}||_V \leq \text{tol}_\text{u}$, where tol_u is a user-supplied tolerance, e.g., $\text{tol}_\text{u} = 1$. Here,

$$\begin{aligned} q(i) &= [q_1(i), \dots, q_d(i)] \in \mathbb{R}^d, \\ ||\vec{Q}||_V &= \max \{ \|q(i)\|_{\mathbb{R}^d}, \quad i_\ell = 1, \dots, n_\ell, \quad \ell = 1, \dots, d \} . \end{aligned} \tag{10.7}$$

If, in contrast, the transformation is not admissible, we perform a regridding and the regridding counter is increased. The displacements of the previous regridding steps with respect to the actual deformation $\vec{W}^{(k)}$ are updated by $\vec{U}^{(k)}$ and denoted by $\vec{Y}^{(\text{regrid\_counter})}$ and then they are set to zero. Note that setting the displacements to zero removes the inner strain of the body with respect to the grid $\vec{X} - \vec{Y}^{(\text{regrid\_counter})}$.

10.4.2 *Interpolation and regridding*

In the interpolation step, the displacement with respect to the last regridding step $\vec{w}^{(k)}$ on the deformed grid $\vec{x} - \vec{u}^{(k)}$ is computed using an interpolation scheme, e.g., d-linear interpolation; cf., Section 3.1.3. Thus, $\vec{x} - \vec{w}^{(k)}$ are the Lagrange coordinates of the strain-free body. The template is interpolated on the deformed grid $\vec{x} - \vec{w}^{(k)} - \vec{u}^{(k)}$.

Figure 10.1 illustrates an original template T_0, the template T_1 on the grid with respect to the latest regridding step, and the template T_2 on the actual grid. For a fixed particle, let x_j, $j = 0, 1, 2$, denote the Euler coordinates of the particle with respect to the three images. Thus, $T_0(x_0) = T_1(x_1) = T_2(x_2)$. The displacement of the particle is phrased in Euler coordinates by $y(x_1) := x_1 - x_0$.

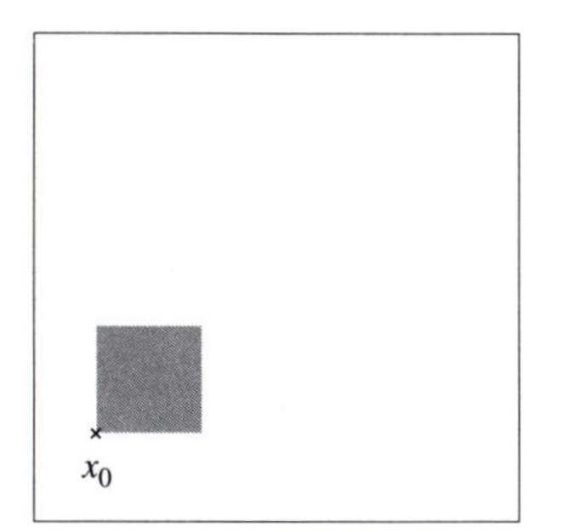

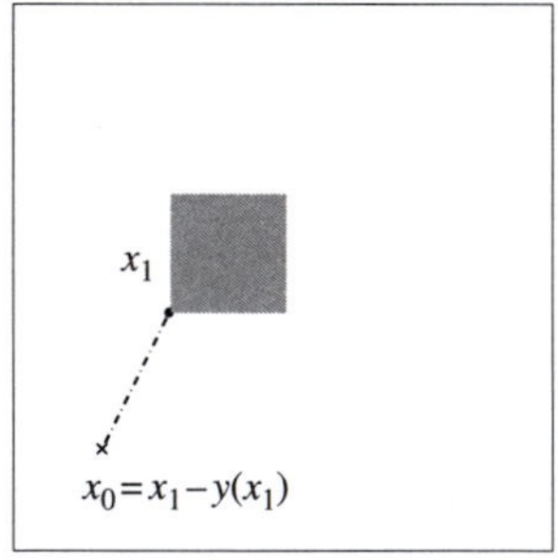

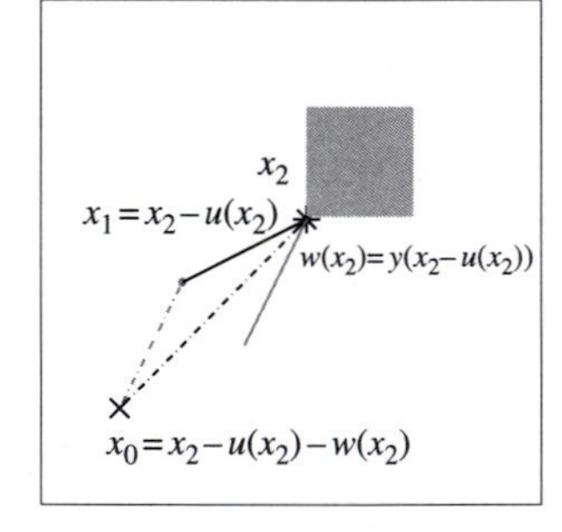

FIG. 10.1 Fluid registration and regridding. LEFT: template T_0 and position of a particle; MIDDLE: template on deformed grid T_1, initial and final position of the particle; RIGHT: template on actual grid T_2, initial, intermediate, and actual position of the particle.

The strain in the body is related to the displacement of x_1 and x_2, which is also phrased in Euler coordinates, $u(x_2) = x_2 - x_1$. The intensity at x_2 is given by

$$\begin{aligned} T_2(x_2) &= T_1(x_2 - u(x_2)) \\ &= T_0(x_1 - y(x_1)) = T_0(x_2 - u(x_2) - y(x_1)) \\ &= T_0(x_2 - u(x_2) - w(x_2)), \end{aligned}$$

where $w(x_2) = y(x_1) = y(x_2 - u(x_2))$. By introducing w, the intermediate T_1 becomes redundant.

10.4.3 *Computing forces and velocities*

For the computation of the forces, we proceed as explained in Section 8.3. Note that in the computation of the finite difference approximations of the gradients the particular boundary conditions enter into play. For the computation of the velocities, we exploit the scheme developed in Section 9.5.

10.4.4 *Stopping criteria*

Different conditions are used as a stopping criterion.

1. As a safeguard, we provide a maximum number of iteration steps.
2. If the relative change of the distance measure is brought below a user-supplied tolerance, e.g., 10^{-5}, the iteration is stopped,

$$\frac{||R - T^{(k-1)}||_F - ||R - T^{(k)}||_F}{||R - T^{(k)}||_F} \leq \text{tol}_{\text{dist}}.$$

   Note that in our implementation a possible division by zero is avoided.
3. The iteration might be stopped if the forces are too small, $||\vec{f}^{(k)}||_V \leq \text{tol}_{\text{for}}$. We do not use this criterion in our implementation.
4. Finally, the iteration might be stopped if the chance of the displacement is too small. Though the implementation of this criterion is straightforward, a description is lengthy if regridding is taken into account. For this reason, this criterion is not contained in the implementation illustrated in Algorithm 10.2.

10.5 Fluid registration: • to C

We continue the registration of a • to a C, see Section 9.8. In Fig. 10.2, we show intermediate results of the fluid registration for $k = 0(40)200$. In order to visualize the deformation, the template has been shaded. Note that the computations have been performed on the original images. In this example, the Lamé constants are $\lambda = 0$ and $\mu = 5000$. The regridding tolerance is 0.025.

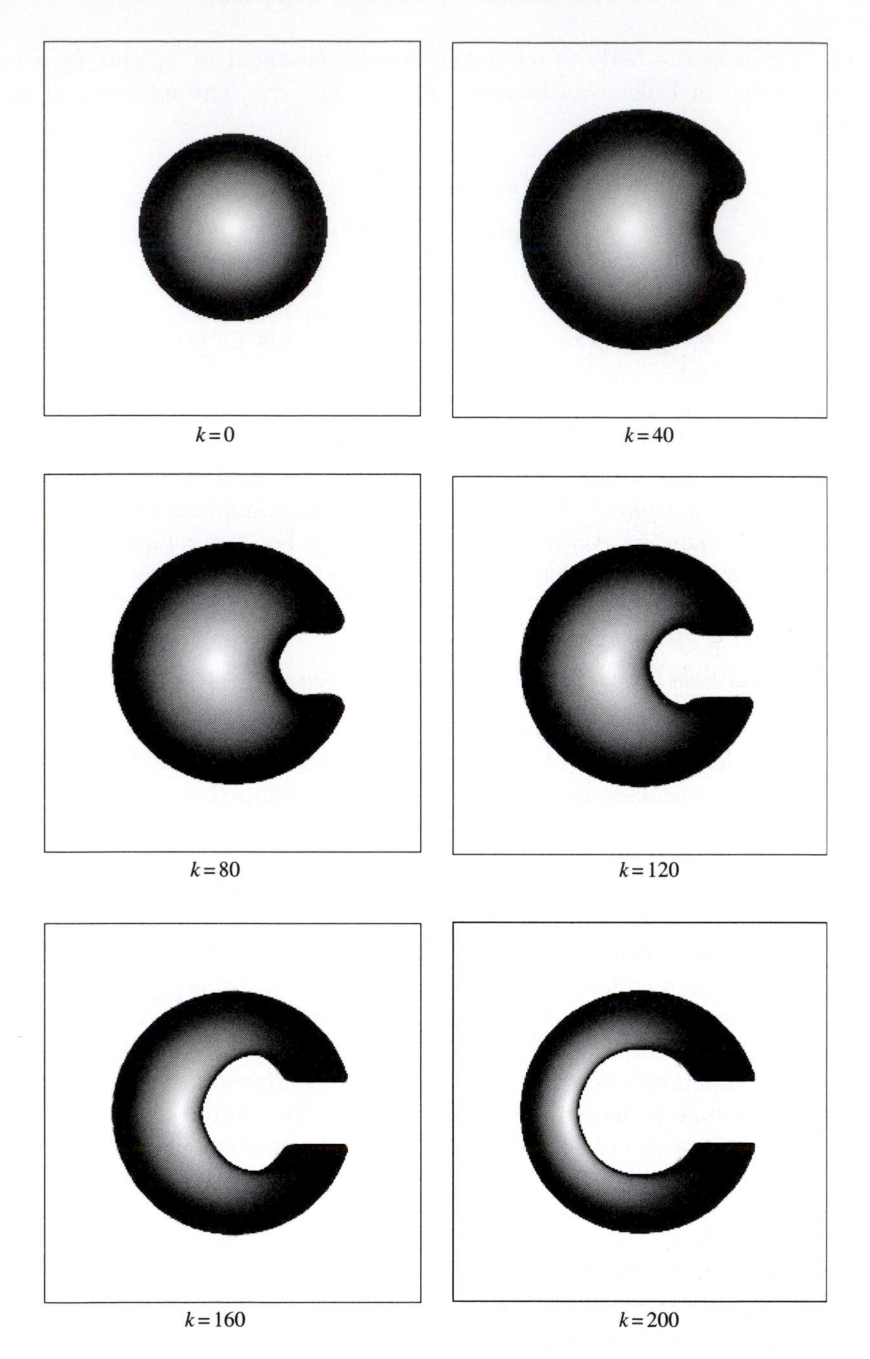

FIG. 10.2 Fluid registration of • to C, $\lambda = 0$, $\mu = 5000$; intermediate results $k = 0(40)200$.

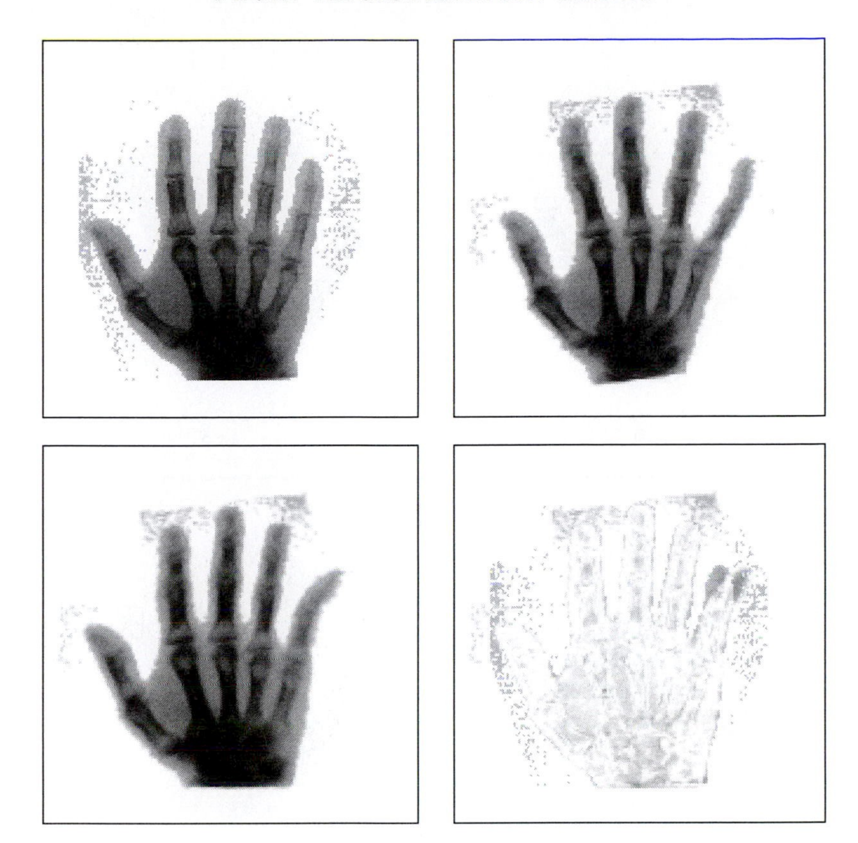

FIG. 10.3 Fluid registration of hands ($\mu = 500$ and $\lambda = 0$), affine linear pre-registered; TOP LEFT: reference image R, TOP RIGHT: template image T, BOTTOM LEFT: fluid registered template T^{fluid} with deformation, and BOTTOM RIGHT: difference $|R - T^{\text{fluid}}|$.

10.6 Fluid registration: hands

Figure 10.3 shows the two modified X-ray images of human hands and the fluid registered hand with an illustration of the deformation, see also Section 9.9. As is apparent from these figures, the fluid registration gives optically pleasing results; see also Fig. 10.4 for intermediate results.

Note that fluid as well as elastic registrations are not able to perform a rigid registration. In the modeling of the strain tensor, we assume that the deformation has no rigid components. We illustrate the importance of this assumption with the registration of the images shown in Fig. 10.5, see also Fig. 10.6. Here, the initial template image has been rotated by 20 degrees.

Though we can see a rotation of the grid by about 20 degrees as well, parts of the image like, for example, the finger tips are deformed in a physically unpleasing fashion.

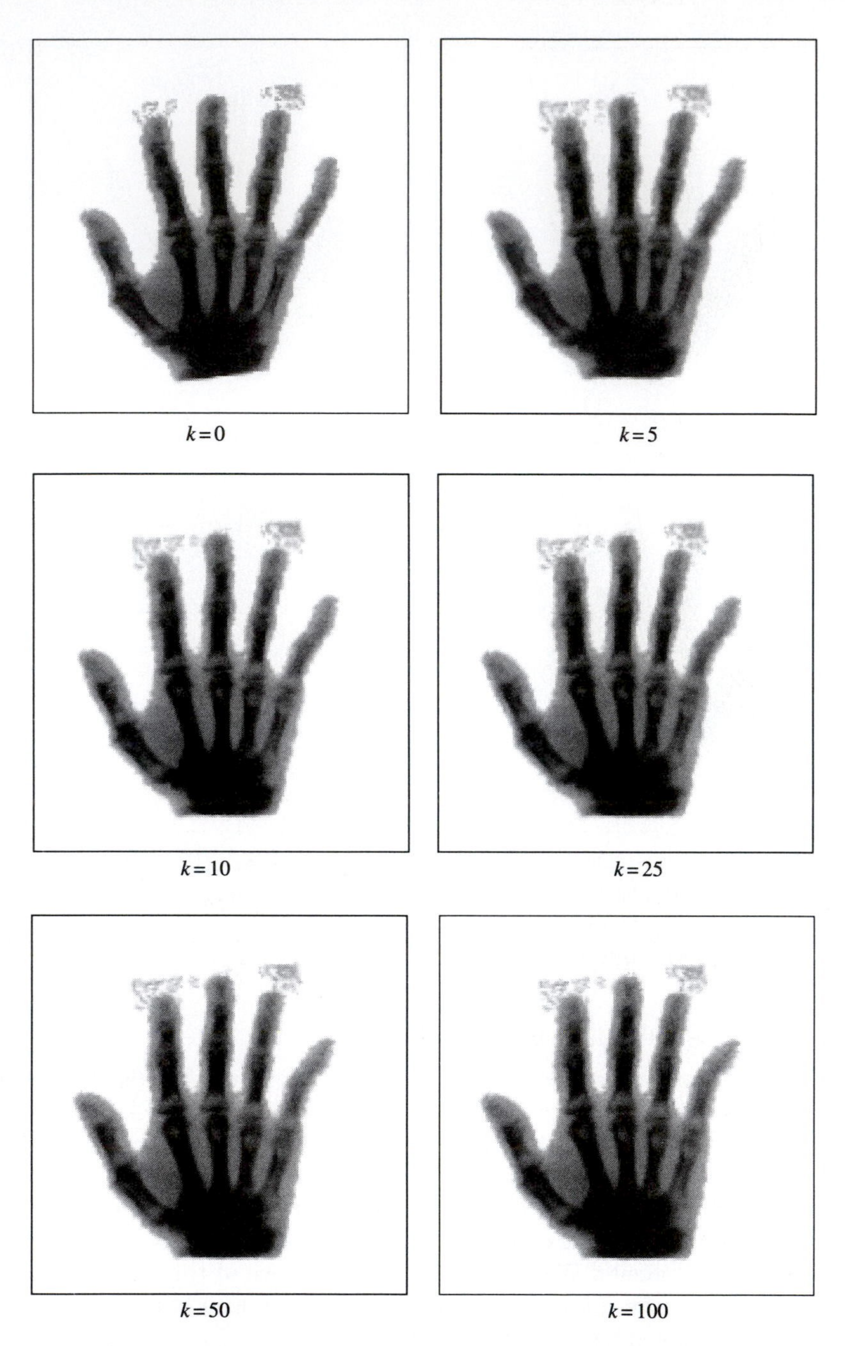

FIG. 10.4 Fluid registration of hands ($\mu = 500$ and $\lambda = 0$), intermediate results for $k = 0, 5, 10, 25, 50, 100$.

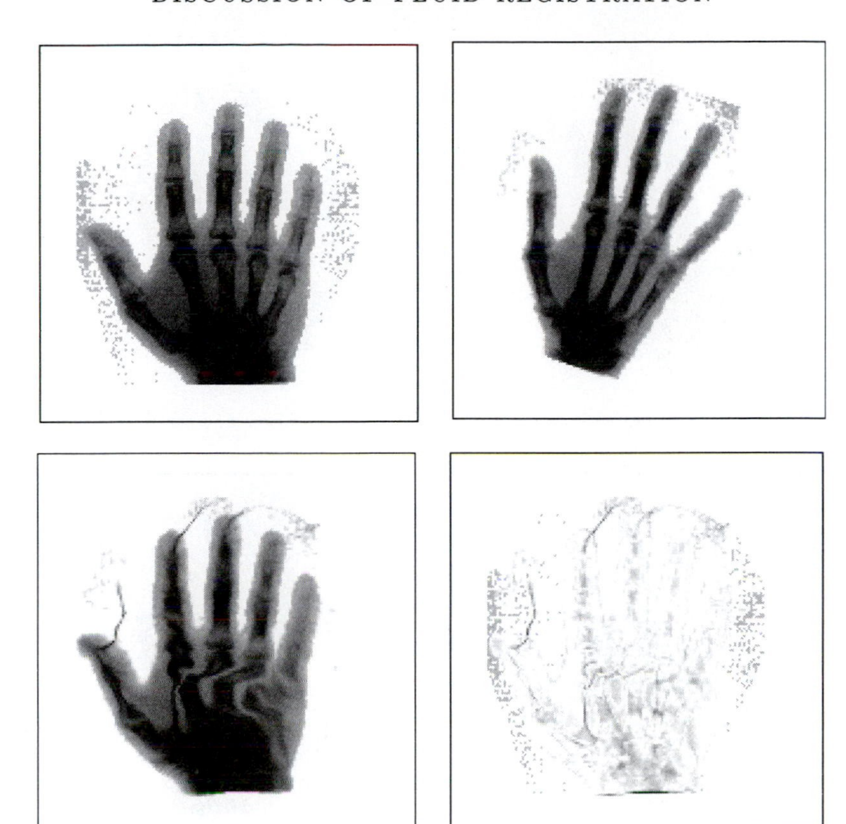

FIG. 10.5 Fluid registration of hands ($\mu = 100$ and $\lambda = 0$), no pre-registration; TOP LEFT: reference image R, TOP RIGHT: template image T, BOTTOM LEFT: fluid registered template T^{fluid} with deformation, and BOTTOM RIGHT: difference $|R - T^{\text{fluid}}|$.

10.7 Fluid registration: hand to disk

In this example, we register the template, a modified version of a hand image from Amit (1994), to the reference image depicted in Fig. 10.7. The images are manipulated such that they share the same gray value range. Besides interpolation artifacts, a complete registration of the template to the reference should be possible. As this example shows, this registration is in fact almost possible. Here, the registration process stopped, because the forces became too small due to interpolation and discretization artifacts.

10.8 Discussion of fluid registration

The fluid registration technique provides a powerful tool for image registration. In principle, it is possible to deform any template image to any reference image,

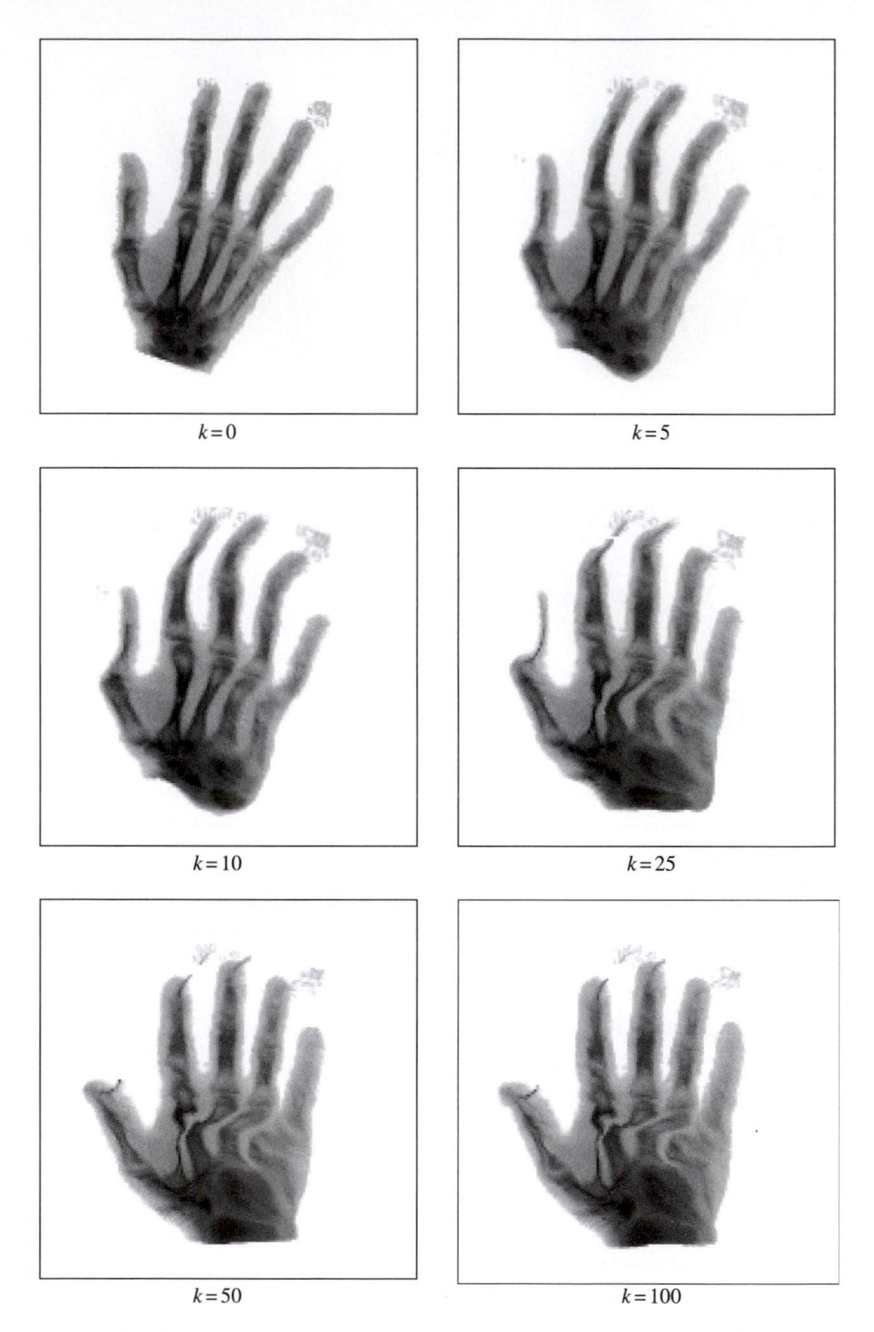

FIG. 10.6 Fluid registration of hands without pre-registration ($\mu = 100$ and $\lambda = 0$); intermediate results for $k = 0, 5, 10, 25, 50, 100$.

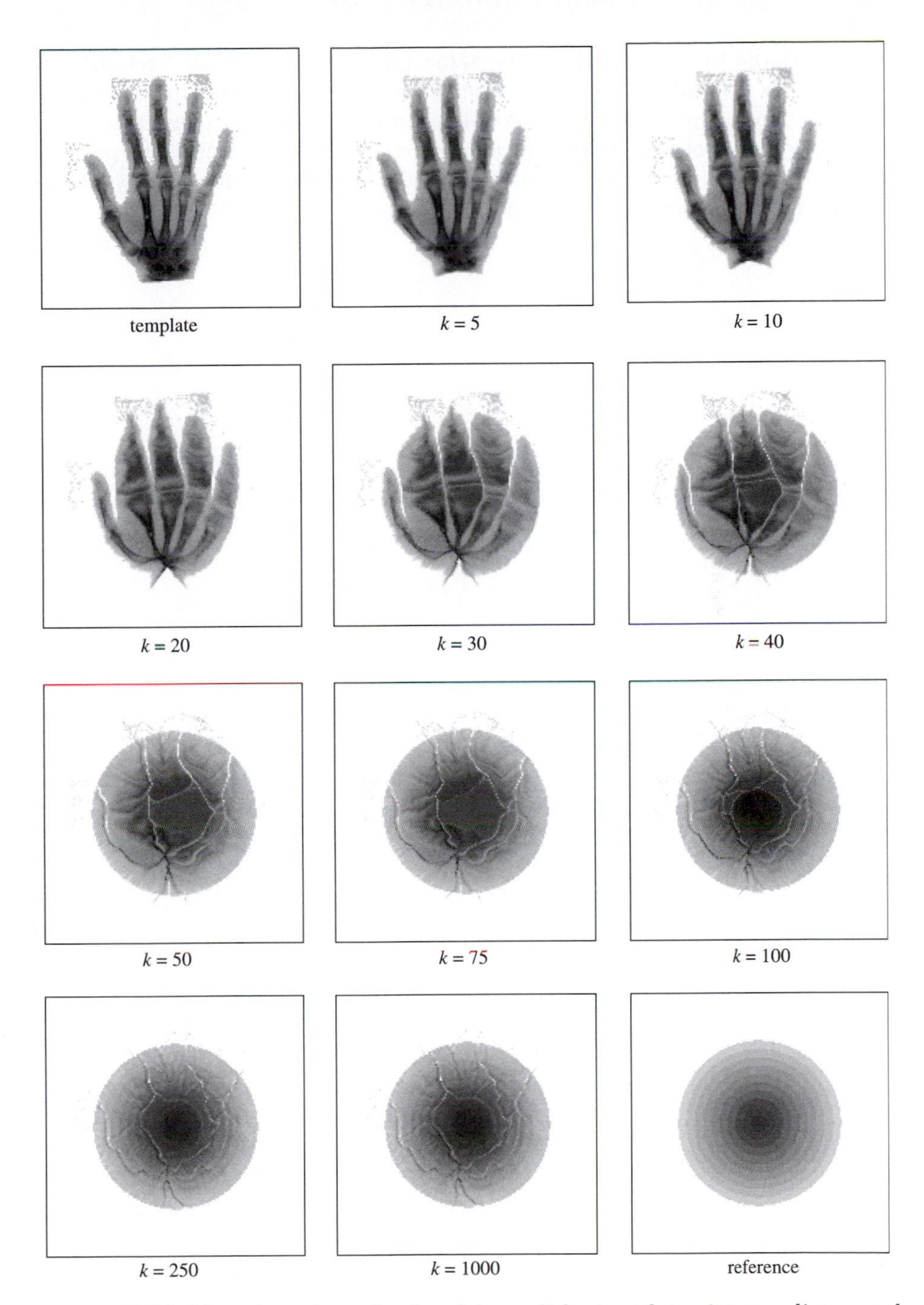

FIG. 10.7 Fluid registration of a hand to a disk, template, intermediate, and reference images.

as long as the two images share the same gray-scale range; see Section 10.7. Though this possibility presents an interesting feature for some applications, it is certainly not appropriate for others. The application of fluid registration for a physically elastic object, as introduced in Section 9.1, is doubtful. The physical motivation of the regularizer is based on fluid-like bodies such as honey. Hands and brains do not deform in general like honey.

11

DIFFUSION REGISTRATION

In this chapter we propose a novel gradient-based regularization term and devise a fast and stable implementation for a finite difference approximation of the underlying partial differential equation. Since this PDE may be viewed as a generalized diffusion equation, we call our new scheme *diffusion registration.*

In contrast to the physically motivated elastic and fluid registrations, see Chapters 9 and 10, the regularizer is motivated by smoothing properties of the displacement. Actually, another important motivation is that a registration step can be performed in $\mathcal{O}(N)$ floating point operations, where N denotes the number of unknowns. The main tool is the so-called additive operator splitting (AOS) scheme; see, e.g., Weickert (1998).

The idea is to split the original problem into a number of simpler problems, which allow for a fast numerical solution. In Section 11.3, we give a new proof for the accuracy of the AOS scheme. The proof is based purely on matrix analysis and therefore the result applies as well to more general situations; cf., Fischer & Modersitzki (1999).

Besides this, we also discuss Thirion's so-called *demons registration*; cf., Thirion (1998). Thirion proposed a method which works well in practice but its derivation is guided by intuition and not entirely understood. In the literature one may find several attempts to shed some light on his approach; see, e.g., Pennec et al (1999) or Bro-Nielsen & Gramkow (1996). Because Thirion's approach offers a variety of possible implementations, the underlying theory is widespread. However, the main point is that he calculates the deformations by regularizing certain driving forces by a Gaussian convolution filter. We show that this technique may be viewed as a special (low-order) approximation to the PDE connected to our new scheme and we thereby gain some new insight into Thirion's approach.

Diffusion registration was introduced by Fischer & Modersitzki (1999) and it is based on the distance measure $\mathcal{D}$ (cf., eqn (8.10)) and the regularizer

$$\mathcal{S}[u] := \frac{1}{2}a[u,u], \quad \text{where} \quad a[u,v] = \sum_{\ell=1}^{d} \int_{\Omega} \langle \nabla u_\ell, \nabla v_\ell \rangle \, dx \tag{11.1}$$

and Neumann boundary conditions are imposed, i.e.,

$$\langle \nabla u_\ell(x), \vec{n}(x) \rangle_{\mathbb{R}^d} = \langle \nabla v_\ell(x), \vec{n}(x) \rangle_{\mathbb{R}^d} = 0 \quad \text{for } x \in \partial\Omega$$

and $\ell = 1, \ldots, d$. Here, $\vec{n}$ denotes the outer normal unit vector of $\partial\Omega$, $\Omega :=]0,1[^d$.

The idea behind this regularizer is to privilege smooth deformations while minimizing oscillations of the components of the displacement, see also Horn & Schunck (1981). Another important advantage of this regularization is that the Euler–Lagrange equations decouple with respect to the spatial directions. As a consequence, the matrix representation A of the finite difference-based discretized derivative of this smoother is a d-by-d block diagonal matrix.

11.1 Continuous and discrete Laplace equations

To begin with, we state the Euler–Lagrange equations for this particular registration.

Theorem 11.1 *The Euler–Lagrange equations for $\mathcal{J} = \mathcal{D} + \alpha\mathcal{S}$, where $\mathcal{D}$ is defined by eqn (8.10) and $\mathcal{S}$ is defined by eqn (11.1) are*

$$f(x, u(x)) + \alpha\Delta u(x) = 0, \quad x \in \Omega, \tag{11.2}$$

$\langle \nabla u_\ell(x), \vec{n}(x)\rangle_{\mathbb{R}^d} = 0$ *for* $x \in \partial\Omega$ *and* $\ell = 1, \ldots, d$.

Proof Follows from Theorem 8.1, $d\mathcal{S}[u; v] = a[u, v]$, and, for a fixed ℓ, setting $\tilde{a}[u_\ell, v_\ell] := \int_\Omega \langle \nabla u_\ell, \nabla v_\ell\rangle \, dx$, from an application of Green's formula,

$$\begin{aligned}
\tilde{a}[u_\ell, v_\ell] &= \int_\Omega \langle \nabla u_\ell, \nabla v_\ell\rangle \, dx \\
&= \int_{\partial\Omega} v_\ell \langle \nabla u_\ell, \vec{n}\rangle \, dA - \int_\Omega (\Delta u_\ell) v_\ell \, dx \\
&= \int_\Omega (\Delta u_\ell) v_\ell \, dx.
\end{aligned}$$

□

For a discretization of the Laplace operator we use the finite difference approximation introduced in Section 8.4 (eqn (3.4)),

$$\Delta u_\ell(X) \approx S^{\mathrm{diff},d} * u_\ell(X). \tag{11.3}$$

For the most important cases $d = 2, 3$ the stencils $S^{\mathrm{diff},d}$ are summarized in Table 11.1. Thus, using the grid notation introduced in Section 8.4 (cf., Definition 3.2), we may also write

$$\Delta[u_\ell](X) \approx S^{\mathrm{diff},d} * u_\ell(X) \quad \text{or} \quad \Delta[u_\ell](\vec{X}) \approx A^{\mathrm{diff},d}\, u_\ell(\vec{X}).$$

Note that the Neumann boundary conditions have to be taken into account; cf. Section 8.4. A discrete formulation of the Euler–Lagrange equations (cf., Theorem 11.1), then reads

$$f(\vec{X}, \vec{U}) + \alpha I_d \otimes A^{\mathrm{diff},d}\, \vec{U} = 0. \tag{11.4}$$

Table 11.1 *Matrix stencils for the discrete Laplace operator.*

$$S^{\text{diff},d}_{n_1,\ldots,n_2} = \begin{cases} -2d, & n_\ell = 2, \ \ell = 1,\ldots,d \\ 1, & n_j = 1,3, n_\ell = 2, \ \ell = 1,\ldots,d, \ \ell \neq j. \end{cases}$$

$$S^{\text{diff},2} = \begin{pmatrix} 0 & 1 & 0 \\ 1 & -4 & 1 \\ 0 & 1 & 0 \end{pmatrix}, \qquad S^{\text{diff},3}_{:,:,1} = \begin{pmatrix} 0 & 0 & 0 \\ 0 & 1 & 0 \\ 0 & 0 & 0 \end{pmatrix},$$

$$S^{\text{diff},3}_{:,:,2} = \begin{pmatrix} 0 & 1 & 0 \\ 1 & -6 & 1 \\ 0 & 1 & 0 \end{pmatrix}, \qquad S^{\text{diff},3}_{:,:,3} = \begin{pmatrix} 0 & 0 & 0 \\ 0 & 1 & 0 \\ 0 & 0 & 0 \end{pmatrix}.$$

11.2 Factorization of the discrete Laplace operator

The matrix representation of the convolution with S, where S is an arbitrary symmetric, d-dimensional stencil, can be expressed recursively. An important building block is the matrix

$$M_m := \begin{pmatrix} 1 & 1 & & & \\ 1 & 0 & 1 & & \\ & \ddots & \ddots & \ddots & \\ & & 1 & 0 & 1 \\ & & & 1 & 1 \end{pmatrix} \in \mathbb{R}^{m \times m}. \tag{11.5}$$

To explain the following complex notation, we present an example for dimension $d = 2$.

Example 11.1 Let $d = 2$ and S be a symmetric matrix stencil,

$$S = \begin{pmatrix} S_{1,1} & S_{2,1} & S_{1,1} \\ S_{1,2} & S_{2,2} & S_{1,2} \\ S_{1,1} & S_{2,1} & S_{1,1} \end{pmatrix}.$$

For the convolution

$$\begin{aligned} v(x_i) &:= S * w(x_i) \\ &= \begin{cases} S_{1,1} w(x_{i_1-1,i_2-1}) + S_{2,1} w(x_{i_1-1,i_2}) + S_{1,1} w(x_{i_1-1,i_2+1}) \\ +S_{1,2} w(x_{i_1,i_2-1}) + S_{2,2} w(x_{i_1,i_2}) + S_{1,2} w(x_{i_1,i_2+1}) \\ +S_{1,1} w(x_{i_1+1,i_2-1}) + S_{2,1} w(x_{i_1+1,i_2}) + S_{1,1} w(x_{i_1+1,i_2+1}) \end{cases} \end{aligned}$$

with respect to Neumann boundary conditions, we may also write

$$\vec{V} = A^{(2)} \cdot \vec{U},$$

where

$$A^{(2)} = \begin{pmatrix} A_2^{(2,1)} + A_1^{(2,1)} & A_1^{(2,1)} & & & \\ A_1^{(2,1)} & A_2^{(2,1)} & \ddots & & \\ & \ddots & \ddots & \ddots & \\ & & \ddots & A_2^{(2,1)} & A_1^{(2,1)} \\ & & & A_1^{(2,1)} & A_2^{(2,1)} + A_1^{(2,1)} \end{pmatrix} \in \mathbb{R}^{N \times N},$$

$N = n_1 n_2$, and the matrices $A_1^{(2,1)}$ and $A_2^{(2,1)}$ are given by

$$\begin{aligned} A_q^{(2,1)} &= S_{2,q} I_{n_1} + S_{1,q} M_{n_1} \\ &= \begin{pmatrix} S_{2,q} + S_{1,q} & S_{1,q} & & & \\ S_{1,q} & S_{2,q} & \ddots & & \\ & \ddots & \ddots & \ddots & \\ & & S_{1,q} & S_{2,q} & S_{1,q} \\ & & & S_{1,q} & S_{2,q} + S_{1,q} \end{pmatrix} \in \mathbb{R}^{n_1 \times n_1}. \end{aligned}$$

Note that Neumann boundary conditions have been incorporated.

Using the Kronecker calculus, we may also write

$$A^{(2)} = I_{n_2} \otimes A_2^{(2,1)} + M_{n_2} \otimes A_1^{(2,1)}.$$

It turns out that the above construction can be applied to any dimension d. For $p_k = 1, 2$, $N_k := n_1 \cdots n_k$, and $k = 1, \ldots, d-1$, we have

$$\begin{aligned} A_{p_{d-1},\ldots,p_1}^{(d,d-1)} &:= I_{n_1} \otimes S_{2,p_{d-1},\ldots,p_1} + M_{n_1} \otimes S_{1,p_{d-1},\ldots,p_1} \\ &\in \mathbb{R}^{n_1 \times n_1}, \end{aligned} \tag{11.6}$$

$$\begin{aligned} A_{p_{d-k},\ldots,p_1}^{(d,d-k)} &:= I_{n_k} \otimes A_{2,p_{d-k-1},\ldots,p_1}^{(d,d-k+1)} + M_{n_k} \otimes A_{1,p_{d-k-1},\ldots,p_1}^{(d,d-k+1)} \\ &\in \mathbb{R}^{N_k \times N_k}, \end{aligned} \tag{11.7}$$

$$\begin{aligned} A^{(d)} &:= I_{n_d} \otimes A_2^{(d,1)} + M_{n_d} \otimes A_1^{(d,1)} \\ &\in \mathbb{R}^{N_d \times N_d}. \end{aligned} \tag{11.8}$$

In particular, for dimension $d = 2, 3$, we have

$$\begin{aligned} A_p^{(2,1)} &= I_{n_1} \otimes S_{2,p} + M_{n_1} \otimes S_{1,p}, \quad p = 1, 2, \\ A^{(2)} &= I_{n_2} \otimes A_2^{(2,1)} + M_{n_2} \otimes A_1^{(2,1)} \end{aligned}$$

(cf., Example 11.1) and

$$\begin{aligned}
A^{(3,2)}_{p,q} &= I_{n_1} \otimes S_{2,p,q} + M_{n_1} \otimes S_{1,p,q}, \quad p,q = 1,2 \\
A^{(3,1)}_{q} &= I_{n_2} \otimes A^{(3,2)}_{2,q} + M_{n_2} \otimes A^{(3,2)}_{1,q}, \quad q = 1,2 \\
A^{(3)} &= I_{n_3} \otimes A^{(3,1)}_{2} + M_{n_3} \otimes A^{(3,1)}_{1}.
\end{aligned}$$

Note that the matrix $-2I_m + M_m$ might be viewed as a discrete version of $\partial_{x,x}$ with Neumann boundary conditions.

If, in particular, the stencils defined in Table 11.1 are used, the description simplifies considerably. Here,

$$\begin{aligned}
A^{\text{diff},2} &= S^{\text{diff},2}_{2,2} I_{n_1 n_2} + S^{\text{diff},2}_{1,2} I_{n_2} \otimes M_{n_1} + S^{\text{diff},2}_{2,1} M_{n_2} \otimes I_{n_1} \\
&= -4I_{n_1 n_2} + I_{n_2} \otimes M_{n_1} + M_{n_2} \otimes I_{n_1}, \\
A^{\text{diff},3} &= S^{\text{diff},3}_{2,2,2} I_{n_1 n_2 n_3} + S^{\text{diff},3}_{1,2,2} I_{n_3} \otimes I_{n_2} \otimes M_{n_1} \\
&\quad + S^{\text{diff},3}_{2,1,2} I_{n_3} \otimes M_{n_2} \otimes I_{n_1} + S^{\text{diff},3}_{2,2,1} M_{n_3} \otimes I_{n_1} \otimes I_{n_1} \\
&= -6I_{n_1 n_2 n_3} + I_{n_3} \otimes I_{n_2} \otimes M_{n_1} \\
&\quad + I_{n_3} \otimes M_{n_2} \otimes I_{n_1} + M_{n_3} \otimes I_{n_1} \otimes I_{n_1}.
\end{aligned}$$

We are now ready to give an eigenvalue decomposition of the matrix $A^{(d)}$. The proof is based on the factorization of M_m and the recursive structure of $A^{(d)}$. For later usage, we introduce

$$C_m := \left(\cos \frac{(2j+1)k\pi}{2m} \right)_{j,k=0,\dots,m-1} \in \mathbb{R}^{m\times m}, \tag{11.9}$$

$$V_m := C_m \operatorname{diag}(\sqrt{1/m}, \sqrt{2/m}, \dots, \sqrt{2/m}) \in \mathbb{R}^{m\times m}, \tag{11.10}$$

$$D_m := 2 \operatorname{diag}\Big(\cos \frac{k\pi}{m},\ k = 0,\dots,m-1 \Big) \in \mathbb{R}^{m\times m}. \tag{11.11}$$

Lemma 11.2 *Let $M_m \in \mathbb{R}^{m\times m}$ be as in eqn (11.5) and C_m, V_m, and D_m be as in eqns (11.9), (11.10), and (11.11), respectively. Then $M_m V_m = V_m D_m$ and $V_m^\top V_m = I_m$.*

Proof Let $k \in \{0,\dots,m-1\}$ be fixed and let $v := C_m e_{k+1}$ be the $(k+1)^{\text{th}}$ column of C_m. From

$$M_m v = (v_1 + v_2, v_1 + v_3, \dots, v_{m-2} + v_m, v_{m-1} + v_m)^\top,$$

and the addition theorem for trigonometric functions,

$$2\cos\alpha\cos\beta = \cos(\alpha - \beta) + \cos(\alpha + \beta),$$

we have

$$\begin{aligned}
v_1 + v_2 &= \cos\frac{k\pi}{2m} + \cos\frac{3k\pi}{2m} = 2\cos\frac{k\pi}{m}\cos\frac{k\pi}{2m} = \lambda_k v_1,\\
v_j + v_{j+2} &= \cos\left(\frac{(2j+1)k\pi}{2m} - \frac{k\pi}{m}\right) + \cos\left(\frac{(2j+1)k\pi}{2m} + \frac{k\pi}{m}\right)\\
&= 2\cos\frac{(2j+1)k\pi}{2m}\cos\frac{k\pi}{m} = \lambda_k v_{j+1}, \quad j = 1,\dots,m-2,\\
v_{m-1} + v_m &= \cos\frac{(2m-3)k\pi}{2m} + \cos\frac{(2m-1)k\pi}{2m} = 2\cos\frac{(m-1)k\pi}{m}\cos\frac{k\pi}{2m}\\
&= 2\cos\frac{k\pi}{m}\cos(k\pi)\cos\frac{k\pi}{2m} = 2\cos\frac{k\pi}{m}\cos\frac{(2m-1)k\pi}{2m} = \lambda_k v_m,
\end{aligned}$$

and thus $M_m C_m = C_m D_m$. Since $M_m = M_m^\top$, the columns of C_m are orthogonal. With $\omega := e^{ik\pi/m}$,

$$\begin{aligned}
e_{k+1}^\top C_m^\top C_m e_{k+1} &= \sum_{j=0}^{m-1} \cos^2\frac{(2j-1)k\pi}{2m} = m + \sum_{j=0}^{m-1} \cos\frac{(2j-1)k\pi}{m}\\
&= m + \sum_{j=0}^{m-1} \Re[e^{ik\pi(2j-1)/m}] = m + \omega \sum_{j=0}^{m-1} \omega^{2j}\\
&= \begin{cases} m, & \omega = 1\\ 2m, & \omega \neq 1, \end{cases}
\end{aligned}$$

where, for $k \neq 0$ and thus $\omega \neq 1$, we used $\sum_{j=0}^{m-1} \omega^{2j} = (\omega^{2m} - 1)/(\omega - 1) = 0$. Hence, V_m is orthonormal. □

The next theorem states that the matrix $A^{(d)}$ (cf., eqn (11.8)) can be diagonalized in terms of discrete cosine transformations (DCTs). It is this property which enables an $\mathcal{O}(N \log N)$ implementation based on a fast DCT.

Theorem 11.3 *Let S be a d-dimensional, symmetric matrix stencil and $A^{(d)}$ be the matrix representation of the convolution with S with respect to Neumann boundary conditions. Then*

$$\begin{aligned}
D^{(d)} &:= (V_{n_d} \otimes \cdots \otimes V_{n_1})^\top A^{(d)} (V_{n_d} \otimes \cdots \otimes V_{n_1})\\
&= \operatorname{diag}(d_{j_1,\dots,j_d},\ j_q = 1,\dots,n_1,\ q = 1,\dots,d),
\end{aligned}$$

where V_m *is defined by eqn (11.10) and* $d_{j_1,\dots,j_d}$ *is defined recursively by*

$$d_{j_1}^{p_{d-1},\dots,p_1} = S_{2,p_{d-1},\dots,p_1} + 2S_{2,p_{d-1},\dots,p_1}\cos(j_1\pi/n_1),$$
$$d_{j_1,\dots,j_k}^{p_{d-k},\dots,p_1} = d_{j_1,\dots,j_k}^{2,p_{d-k},\dots,p_1} + 2d_{j_1,\dots,j_k}^{1,p_{d-k},\dots,p_1}\cos(j_k\pi/n_k),$$
$$d_{j_1,\dots,j_d} = d_{j_1,\dots,j_{d-1}}^{2} + 2d_{j_1,\dots,j_{d-1}}^{1}\cos(j_d\pi/n_d).$$

Proof From Lemma 11.2, we have for $A_{p_{d-1},\dots,p_1}^{(d,d-1)}$ (cf., eqn (11.6))

$$D_{p_{d-1},\dots,p_1}^{(d,d-1)} := V_{n_1}^\top A_{p_{d-1},\dots,p_1}^{(d,d-1)} V_{n_1} = \operatorname{diag}\left(d_{j_1}^{p_{d-1},\dots,p_1},\ j_1 = 1,\dots,n_1\right),$$

where $d_{j_1}^{p_{d-1},\dots,p_1} = S_{2,p_{d-1},\dots,p_1} + 2S_{2,p_{d-1},\dots,p_1}\cos((j_1-1)\pi/n_1)$.

Using the recurrence formula (11.7), we have

$$D_{p_{d-k},\dots,p_1}^{(d,d-k)} := (V_{n_k}\otimes\cdots\otimes V_{n_1})^\top A_{p_{d-k},\dots,p_1}^{(d,d-k)} (V_{n_k}\otimes\cdots\otimes V_{n_1}) = \operatorname{diag}\left(d_{j_1,\dots,j_k}^{p_{d-k},\dots,p_1},\ j_q = 1,\dots,n_q,\ q = 1,\dots,k\right),$$

where $d_{j_1,\dots,j_k}^{p_{d-k},\dots,p_1} = d_{j_1,\dots,j_k}^{2,p_{d-k},\dots,p_1} + 2d_{j_1,\dots,j_k}^{1,p_{d-k},\dots,p_1}\cos((j_k-1)\pi/n_k)$.

Finally, from eqn (11.6),

$$D^{(d)} := (V_{n_d}\otimes\cdots\otimes V_{n_1})^\top A^{(d)} (V_{n_d}\otimes\cdots\otimes V_{n_1}) = \operatorname{diag}(d_{j_1,\dots,j_d},\ j_q = 1,\dots,n_q,\ q = 1,\dots,d)$$

where $d_{j_1,\dots,j_d} = d_{j_1,\dots,j_{d-1}}^{2} + 2d_{j_1,\dots,j_{d-1}}^{1}\cos((j_d-1)\pi/n_d)$. □

The following corollaries summarize the statement of Theorem 11.3 for the important cases $d = 2$ and $d = 3$, respectively.

Corollary 11.4 *Let* $A^{\text{diff},2}$ *be the matrix associated with* $S^{\text{diff},2}$*; cf., Table 11.1. Then*

$$D^{\text{diff},2} := (V_{n_2}\otimes V_{n_1})^\top A^{\text{diff},2}(V_{n_2}\otimes V_{n_1}) = \operatorname{diag}(d_{j_1,j_2},\ j_1 = 1,\dots,n_1,\ j_2 = 1,\dots,n_2),$$

where
$d_{j_1,j_2} = -4 + 2\cos((j_1-1)\pi/n_1) + 2\cos((j_2-1)\pi/n_2)$,
$j_1 = 1,\dots,n_1,\ j_2 = 1,\dots,n_2$.

Proof Follows directly from Theorem 11.3. □

Corollary 11.5 *Let $A^{\mathrm{diff},3}$ be the matrix associated with $S^{\mathrm{diff},3}$; cf., Table 11.1. Then*

$$\begin{aligned} D^{\mathrm{diff},3} &:= (V_{n_3} \otimes V_{n_2} \otimes V_{n_1})^\top A^{\mathrm{diff},3}(V_{n_3} \otimes V_{n_2} \otimes V_{n_1}) \\ &= \mathrm{diag}(d_{j_1,j_2,j_3},\ j_\ell = 1,\dots,n_\ell,\ \ell = 1,2,3), \end{aligned}$$

where

$$\begin{aligned} d_{j_1,j_2,j_3} = &-6 + 2\cos((j_1-1)\pi/n_1) + 2\cos((j_2-1)\pi/n_2) \\ &+ 2\cos((j_3-1)\pi/n_3),\ j_\ell = 1,\dots,n_\ell,\ \ell = 1,2,3. \end{aligned}$$

Proof Follows directly from Theorem 11.3. □

As a consequence of the decomposition stated in Theorem 11.3 and Corollaries 11.4 and 11.5, a fast DCT can be exploited for a numerical solution of the linear system (11.4). Thus, an $\mathcal{O}(N \log N)$ implementation based on real arithmetic can be deduced. However, in Section 11.3 we present an $\mathcal{O}(N)$ implementation of a time marching approach based on an additive operator splitting scheme.

11.3 Additive operator splitting (AOS)

A popular approach to solve a non-linear PDE like eqn (11.2) is to introduce an artificial time t and to compute the steady state solution $\partial_t u(x,t) = 0$ of the time-dependent PDE

$$\partial_t u(x,t) = f(x, u(x,t)) + \alpha\Delta u(x,t), \quad x \in \Omega,\ t \geq 0, \tag{11.12}$$

via a time marching algorithm, see also eqn (8.9). To overcome the non-linearity in f, we employ the following semi-implicit iterative scheme,

$$\partial_t u(x,t_{k+1}) - \alpha\Delta u(x,t_{k+1}) = f(x, u(x,t_k)), \quad k = 0,1,\dots, \tag{11.13}$$

where $u(x,t_0)$ is some initial deformation, typically $u(x,t_0) = 0$. In other words, the trick is to compute the driving force f for the previous solution $u(x,t_k)$ and subsequently to solve for $u(x,t_{k+1})$.

An important property of the system of equations (11.13) is that they are essentially decoupled. The coupling is only through the right hand side. The jth equation reads

$$\partial_t u_j(x,t_{k+1}) - \alpha\Delta u_j(x,t_{k+1}) = f_j(x, u(x,t_k)), \quad k = 0,1,\dots. \tag{11.14}$$

Note that eqn (11.14) is nothing but an inhomogeneous heat equation and well understood; see, e.g., Folland (1995, Th 4.8).

There exist many schemes to solve (11.14) numerically. Here, we are interested in schemes which are on the one hand accurate and stable and on the other hand fast and efficient.

For the time discretization we introduce a time-step $\tau > 0$ and for the spatial discretization the grid $\vec{X}$; cf., Definition 3.2. For a fixed j, we set

$$\vec{V}^{(k)} := u_j(\vec{X}, k\tau) \tag{11.15}$$

and

$$\vec{F}_j^{(k)} := f_j(\vec{X}, u(\vec{X}, \tau k)). \tag{11.16}$$

Using these abbreviations the discrete version of eqn (11.14) reads

$$\frac{\vec{V}^{(k+1)} - \vec{V}^{(k)}}{\tau} - \sum_{\ell=1}^{d} A_\ell \vec{V}^{(k+1)} = \vec{F}_j^{(k)}, \tag{11.17}$$

where

$$\frac{\vec{V}^{(k)} - \vec{V}^{(k-1)}}{\tau} \approx \partial_t u_j(\vec{X}, k\tau)$$

is a forward difference approximation of the time derivative $\partial_t u_j$ with time-step τ and $A_\ell \in \mathbb{R}^{n \times n}$ is an appropriate finite difference approximation of the second order derivative of u_j with respect to the ℓth space coordinate,

$$A_\ell \vec{V}^k \approx \alpha\, \partial_{x_\ell, x_\ell} u_j(\vec{X}, \tau k). \tag{11.18}$$

In the case of image registration, we have chosen a simple three-point star leading to an essentially tridiagonal matrix A_ℓ; cf., Table 11.1. However, in the light of Theorem 11.6, there is no particular assumption on A_ℓ.

After rearranging eqn (11.17), we obtain the semi-implicit scheme for (11.14)

$$\vec{V}_{\text{IS}}^{(k+1)} := \left(I - \tau \sum_{\ell=1}^{d} A_\ell\right)^{-1} \left(\vec{V}_{\text{IS}}^{(k)} + \tau \vec{F}_j^{(k)}\right), \quad k = 0, 1, \ldots \tag{11.19}$$

It is known that this scheme is of order one with respect to the time-step τ, and of order two with respect to the spatial meshsize.

The iteration (11.19) requires the solution of a linear system with n unknowns at each time-step. Note that the systems connected to the individual matrices A_ℓ are essentially tridiagonal and may be solved by an $\mathcal{O}(n)$ direct scheme. On the other hand, the sum is not tridiagonal and therefore the system in (11.19) does not permit such a fast implementation, in general.

The idea of AOS is to replace the inverse of the sum by a sum of inverses. The corresponding iterates are defined by

$$\vec{V}_{\mathrm{AOS}}^{(k+1)} := \frac{1}{d}\sum_{\ell=1}^{d}(I - d\tau A_\ell)^{-1}\left(\vec{V}_{\mathrm{AOS}}^{(k)} + \tau\vec{F}_j^{(k)}\right), \quad k = 0, 1, \dots \tag{11.20}$$

This clever decomposition allows an $\mathcal{O}(n)$ implementation by employing the Thomas algorithm (1949).

11.3.1 *Matrix analysis of AOS*

The next theorem relates the iteration matrices of the semi-implicit and the AOS scheme to each other. It turns out that the distance between these two matrices is surprisingly small if the time-step has been chosen sufficiently small. Note that for image registration, we are not interested in time accuracy and τ may not be small. However, our result is based on matrix analysis, it is not restricted to matrices stemming from PDE discretization, and is therefore of interest independently of these facts. The following theorem is formulated for general matrices. Matrix distances are measured in the spectral norm: for $A \in \mathbb{C}^{n\times n}$,

$$\|A\|_2 := \max\{\sqrt{\lambda} : \ \lambda \text{ is eigenvalue of } A^{\mathrm{H}}A\}.$$

Theorem 11.6 *Let $d \in \mathbb{N}$, $\tau \geq 0$, and let $A_1, \dots, A_d \in \mathbb{R}^{n\times n}$ be simultaneously diagonalizable with eigenvalues in the left half plane. Then there exists a constant $C \in \mathbb{R}$ with*

$$\left\| \left(I - \tau\sum_{\ell=1}^{d} A_\ell\right)^{-1} - \frac{1}{d}\sum_{\ell=1}^{d}(I - d\tau A_\ell)^{-1} \right\|_2 \leq C \cdot \tau^2.$$

Proof The idea is to employ a basis of eigenvectors such that the corresponding matrices become diagonal. To this end consider

$$WA_\ell W^{-1} = \Lambda_\ell = \mathrm{diag}(\lambda_{\ell,j}, \ 1 \leq j \leq n),$$

where W is an eigenvector matrix of any A_ℓ and the Λ_ℓ's are the diagonal matrices based on the individual eigenvalues. Hence,

$$W(I - d\tau A_\ell)W^{-1} = I - d\tau\Lambda_\ell,$$

$$W\left(I - \tau\sum_{\ell=1}^{d} A_\ell\right)W^{-1} = \left(I - \tau\sum_{\ell=1}^{d}\Lambda_\ell\right),$$

and

$$W\left[\left(I-\tau\sum_{\ell=1}^{d}A_\ell\right)^{-1}-\frac{1}{d}\sum_{\ell=1}^{d}(I-d\tau A_\ell)^{-1}\right]W^{-1}$$
$$=\left(I-\tau\sum_{\ell=1}^{d}\Lambda_\ell\right)^{-1}-\frac{1}{d}\sum_{\ell=1}^{d}(I-d\tau\Lambda_\ell)^{-1}$$

is a diagonal matrix, where the kth diagonal entry is given by

$$q_k=\varphi\left(\tau\sum_{\ell=1}^{d}\lambda_{\ell,k}\right)-\frac{1}{d}\sum_{\ell=1}^{d}\varphi(d\tau\lambda_{\ell,k}),\quad \varphi(x)=\frac{1}{1-x}.$$

A Taylor expansion of the analytic function φ at $x_0=0$ reads

$$\varphi(x)=1+x+\frac{2x^2}{(1-\xi)^3},\quad \xi=\xi(x)\in[0,x].$$

This yields

$$\begin{aligned}
q_k &= 1+\tau\sum_{\ell=1}^{d}\lambda_{\ell,k}+\frac{2\tau^2(\sum_{\ell=1}^{d}\lambda_{\ell,k})^2}{(1-\xi)^3}\\
&\quad -\frac{1}{d}\sum_{\ell=1}^{d}\left(1+d\tau\lambda_{\ell,k}+\frac{2(d\tau\lambda_{\ell,k})^2}{(1-\xi_\ell)^3}\right)\\
&= \tau^2\left(\frac{2(\sum_{\ell=1}^{d}\lambda_{\ell,k})^2}{(1-\xi)^3}-\frac{1}{d}\sum_{\ell=1}^{d}\frac{2(d\lambda_{\ell,k})^2}{(1-\xi_\ell)^3}\right)\\
&=: \tau^2 g(\lambda_{1,k},\ldots,\lambda_{d,k}).
\end{aligned}$$

By assumption we can find compact sets Q_ℓ contained in the left complex half plane which enclose all eigenvalues of A_ℓ. Consequently, the function g is continuous on $Q:=Q_1\times\cdots\times Q_d$ and attains its maximum

$$\tilde{C}:=\max\{|g(z)|\ :\ z\in Q\}.$$

We thus have $|q_k|\le\tilde{C}\tau^2$, for $k=1,\ldots,n$, and the statement follows from

$$\begin{aligned}
&\left\|\left(I-\tau\sum_{\ell=1}^{d}A_\ell\right)^{-1}-\frac{1}{d}\sum_{\ell=1}^{d}(I-d\tau A_\ell)^{-1}\right\|_2\\
&\quad\le\|W\|_2\|W^{-1}\|_2\cdot\left\|\left(I-\tau\sum_{\ell=1}^{d}\Lambda_\ell\right)^{-1}-\frac{1}{d}\sum_{\ell=1}^{d}(I-d\tau\Lambda_\ell)^{-1}\right\|_2\le C\cdot\tau^2.
\end{aligned}$$

□

It is worth noticing that matrices are simultaneously diagonalizable if and only if they commute with each other, a proof of which can be found for example in Horn & Johnson (1990, Th. 1.3.19). It is this property which provides a convenient tool for checking the assumption of the above theorem.

In the statement of the theorem it is assumed that all eigenvalues are contained in the left half plane. This assumption ensures that, independent of the value of τ, the matrices

$$I - \tau \sum_{\ell=1}^{d} A_\ell \quad \text{and} \quad I - d\tau A_\ell$$

are non-singular. A close inspection of the proof shows that the theorem holds for arbitrary eigenvalues as well, as long as τ is small enough.

11.3.2 *Diffusion registration using AOS*

Now let us return to the PDE discretization (11.19) and (11.20), respectively. The next corollary relates the solutions of the two time marching processes to each other. In accordance with Theorem 11.6, the statement of the corollary is valid for general matrices.

Corollary 11.7 *Let $d, K \in \mathbb{N}$, $\tau \geq 0$, and let $A_1, \ldots, A_d \in \mathbb{R}^{n\times n}$ be simultaneously diagonalizable with eigenvalues in the left half plane. Moreover, let $\vec{V}_{\mathrm{IS}}^{(k+1)}$ and $\vec{V}_{\mathrm{AOS}}^{(k+1)}$ denote the solution of eqns (11.19) and (11.20), respectively. Then there exists a constant $C > 0$ with*

$$\left\| \vec{V}_{\mathrm{IS}}^{(k+1)} - \vec{V}_{\mathrm{AOS}}^{(k+1)} \right\|_2 \leq C \cdot \tau^2, \quad 0 \leq k \leq K.$$

The particular matrices introduced by (11.18) are given by

$$A_\ell = I \otimes \cdots \otimes I \otimes B_\ell \otimes I \otimes \cdots \otimes I,$$

where the ℓth factor B_ℓ is an approximation of the second order derivative in only one spatial direction and $\otimes$ denotes the Kronecker product of matrices. More precisely, we have

$$B_\ell = \begin{pmatrix} \alpha_{\ell,1} & \beta_{\ell,2} & & 0 \\ \gamma_{\ell,2} & \alpha_{\ell,2} & \ddots & \\ & \ddots & \ddots & \beta_{\ell,m} \\ 0 & & \gamma_{\ell,m} & \alpha_{\ell,m} \end{pmatrix},$$

with appropriate values of $\alpha_{\ell,j}$, $\beta_{\ell,j}$, and $\gamma_{\ell,j}$ satisfying $-\alpha_{\ell,j} \geq |\beta_{\ell,j+1}| + |\gamma_{\ell,j}|$. Using the particular stencils of Table 11.1, the entries become $\alpha_j = -2$, $\beta_j = \gamma_j = 1$.

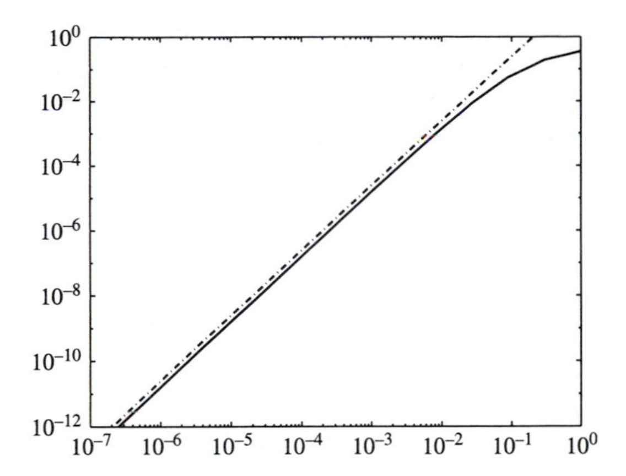

FIG. 11.1 $\|(I - \tau\sum_{\ell=1}^{d} A_\ell)^{-1} - \frac{1}{d}\sum_{\ell=1}^{d}(I - d\tau A_\ell)^{-1}\|_2$ (solid line) and $C \cdot \tau^2$ (dash–dotted line) versus τ, where $C = 25$ for the matrices arising in diffusion registration.

Obviously, the matrices A_ℓ commute and since the B_ℓ have negative eigenvalues, Theorem 11.6 and Corollary 11.7 apply. In other words, the iterates of the semi-implicit scheme and the AOS scheme differ by $\mathcal{O}(\tau^2)$.

It is apparent from the proof of Theorem 11.6 that the constant C depends on the eigenvalues and eigenvectors of the matrices A_ℓ. For the particular matrices arising in diffusion registration, it turns out that C is of moderate size and essentially independent of the number of grid points. To illustrate this fact we have computed the quantity

$$\left\| \left(I - \tau \sum_{\ell=1}^{d} A_\ell \right)^{-1} - \frac{1}{d} \sum_{\ell=1}^{d} (I - d\tau A_\ell)^{-1} \right\|_2$$

for various sizes of A_ℓ and various τ. Figure 11.1 shows a representative result for $d = 2$, $n_1 = n_2 = 1024$, where $C = 25$ serves as an upper bound. The plots for other sizes of A_ℓ are visually indistinguishable from the displayed one.

The overall algorithm for the AOS scheme is summarized in Algorithm 11.8. Note that the main steps in the algorithm are the computation of the force $\vec{F}^{(k)}$ related to the chosen distance measure and the solution of the linear systems related to our particular smoother. We use a standard $\mathcal{O}(n)$ d-linear interpolation technique for computation of the force. As already mentioned, the matrices $I - d\tau A_\ell$ are tridiagonal and strictly diagonally dominant. Hence, the $\mathcal{O}(n)$ Thomas algorithm (1949) is a numerically stable solution technique. In conclusion, we end up with a fast and efficient $\mathcal{O}(n)$ registration algorithm.

Moreover, the implementation offers a coarse grain parallelism based on the ℓ-loop in the algorithm. Due to the special Kronecker product structure of the matrices A_ℓ, a fine grain parallelism can be exploited, too. For example, in two

Algorithm 11.8 *AOS-based diffusion registration for two d-dimensional images R and T.*

Initialize $k = 0$, $\vec{U}^{(k)} = (\vec{U}_1^{(k)}, \dots, \vec{U}_d^{(k)}) = 0$.
For $k = 0, 1, 2, \dots$
  For $j = 1, \dots, d$,
    % Compute the jth component of the force field
    $\vec{F}_j^{(k)} = \left(T(\vec{X} - \vec{U}^{(k)}) - R(\vec{X})\right) \cdot \partial_j T(\vec{X} - \vec{U}^{(k)})$.
    % Compute the AOS iterate
    For $\ell = 1, \dots, d$,
      Solve $(I - d\tau A_\ell)\vec{V}_\ell = \vec{U}_j^{(k)} + \tau \vec{F}_j^{(k)}$.
    End.
    Set $\vec{U}_j^{(k)} = \frac{1}{d} \sum_{\ell=1}^{d} \vec{V}_\ell$.
  End.
End.

dimensions we have to solve the two linear systems

$$((I - 2\tau B_1) \otimes I)\vec{V}_1 = \vec{U}_j^{(k)} + \tau \vec{F}_j^{(k)},$$

$$(I \otimes (I - 2\tau B_2))\vec{V}_2 = \vec{U}_j^{(k)} + \tau \vec{F}_j^{(k)}.$$

Each of these systems decouples into a number of small systems which can be solved independently in parallel.

11.4 Diffusion registration: • to C

We continue the registration of a • to a C; see Section 9.8 and Section 10.5.

In Figs 11.2 and 11.3 we show intermediate results of diffusion registration with $\alpha = 50$ and $\tau = 0.02$ using the AOS- and the DCT-based solvers, see Section 11.2, respectively. Finally, Fig. 11.4 directly compares the registration results with respect to the numerical convergence and the grids. These figures indicate that the registrations based on the AOS and the semi-implicit schemes are indistinguishable – at least for this example.

Note that it is not self-evident that the two methods are so close together. Particularly for large values of τ, the underlying models are different and thus one may expect different qualitative results.

Diffusion registration might be compared with elastic registration; see Chapter 9. In both techniques the registration is smoothed by regularizing the

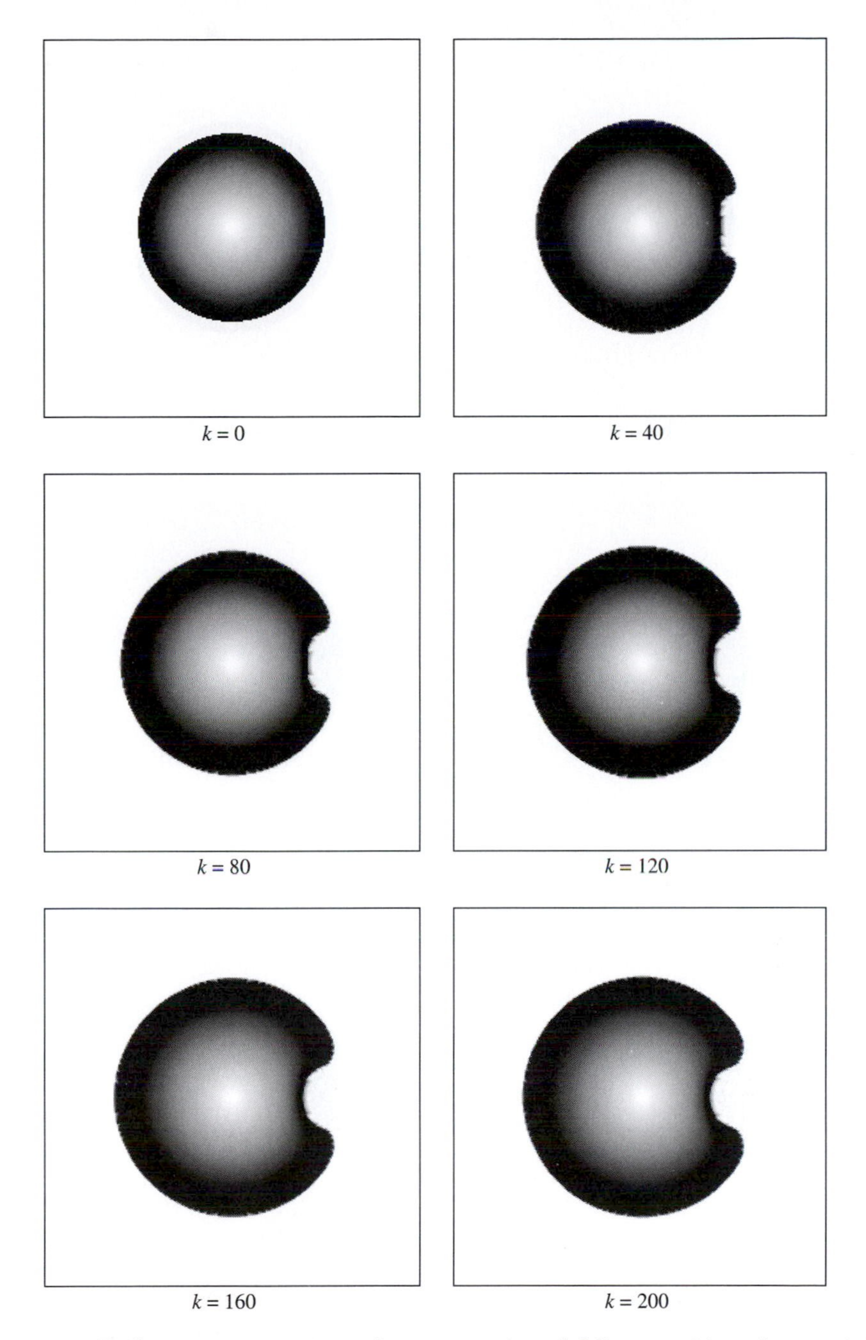

FIG. 11.2 Diffusion registration of • to C using AOS, $\alpha = 50$ and $\tau = 0.02$; intermediate results for $k = 0(40)200$.

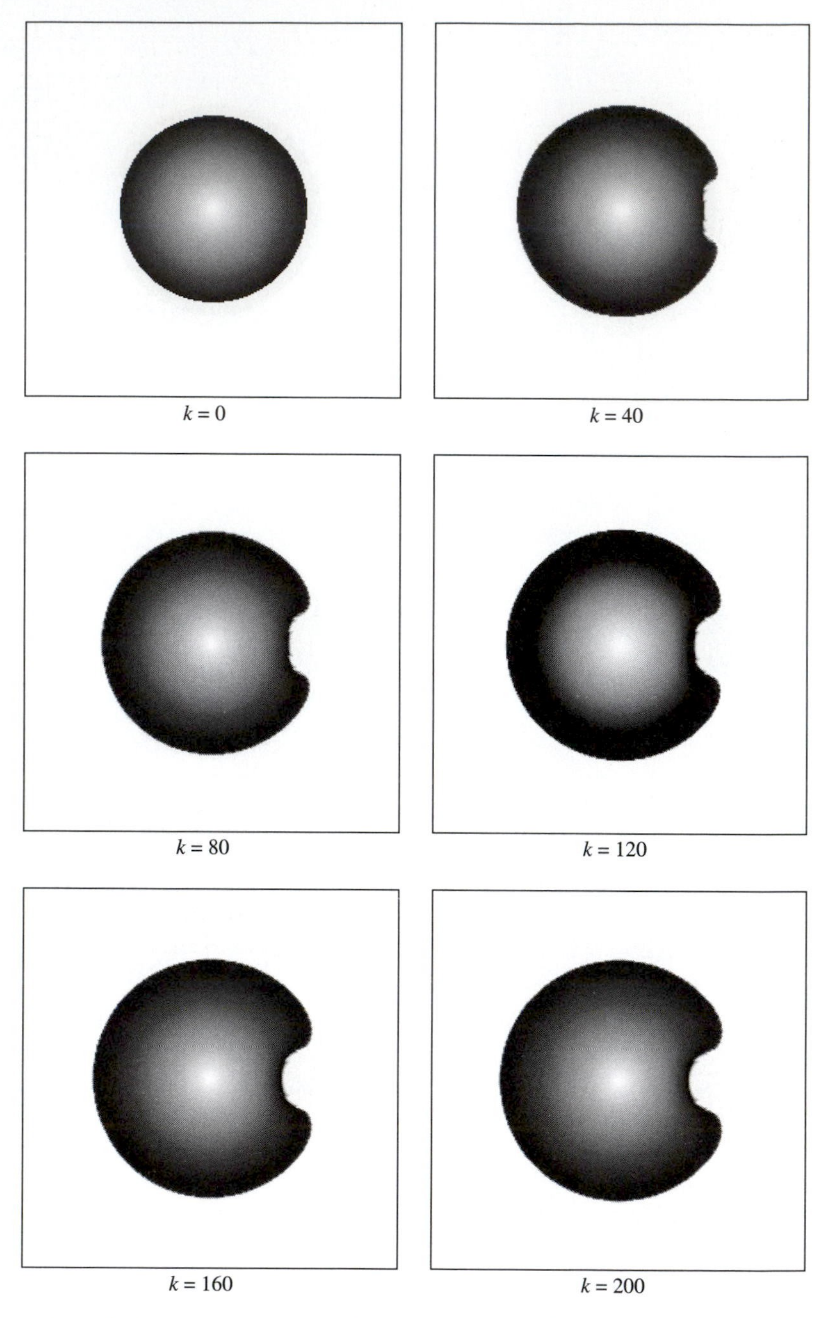

FIG. 11.3 Diffusion registration of • to C using DCT, $\alpha = 50$ and $\tau = 0.02$; intermediate results for $k = 0(40)200$.

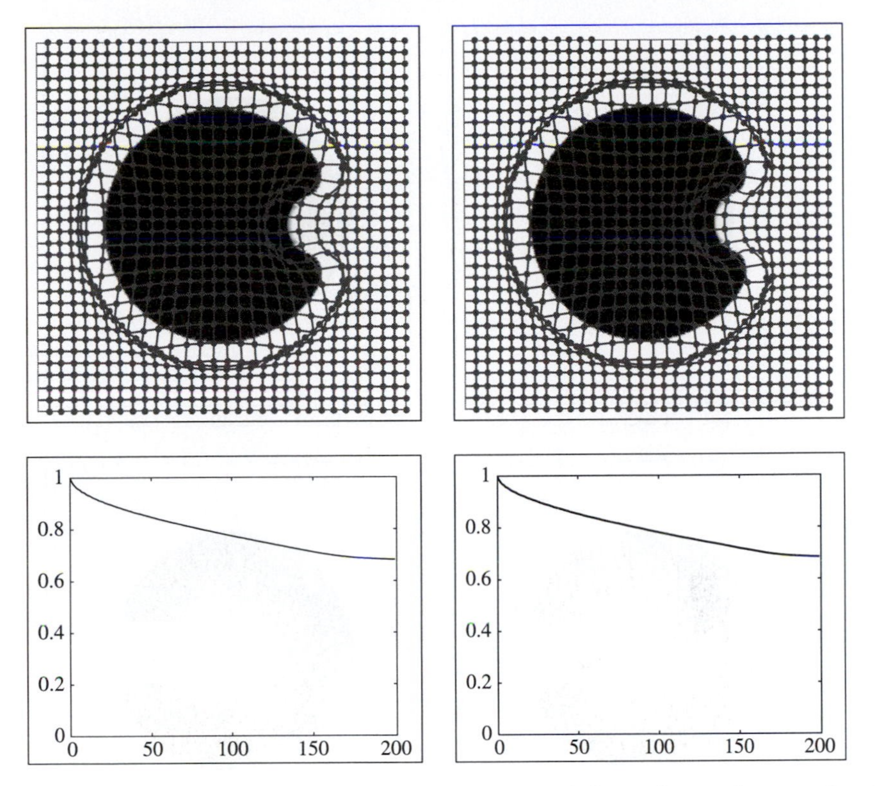

FIG. 11.4 Diffusion registration of • to C. TOP: deformed template and grid; BOTTOM: iteration history; LEFT: AOS-based, RIGHT: DCT-based.

displacement u, whereas in fluid registration (cf., Chapter 10) the regularization is based on the velocity $v = \partial_t u$. In contrast to elastic registration, where the regularization is based on second order derivatives (cf., Definition 9.1), diffusion registration uses first order derivatives.

The extension of our diffusion registration to a *fluid*-type formulation is straightforward. Results are shown in Fig. 11.5.

11.5 Diffusion registration: hands

Figure 11.6 shows the two modified X-ray images of human hands and the diffusion registered hand with an illustration of the deformation; see also Section 9.9 and Section 10.6. In Fig. 11.7 we show the intermediate results for various iterations.

As for the example shown in the previous sections, the results of the AOS-based (Figs 11.6 and 11.7) and the ones from the DCT-based diffusion registration (Figs 11.8 and 11.9) are almost indistinguishable.

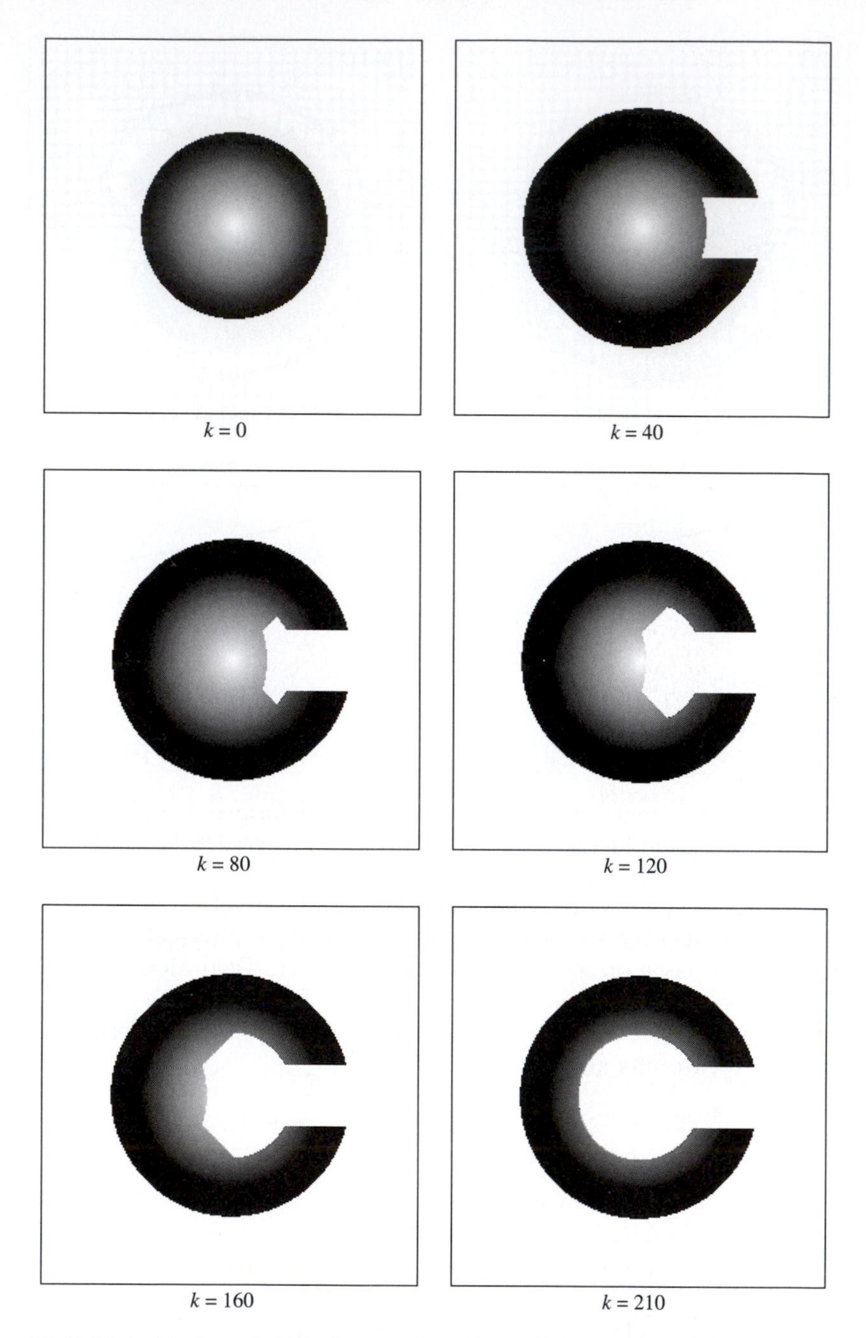

FIG. 11.5 Velocity-based diffusion registration of • to C using the AOS scheme, $\alpha = 1$ and $\tau = 0.2$; intermediate results for $k = 0(40)160, 210$.

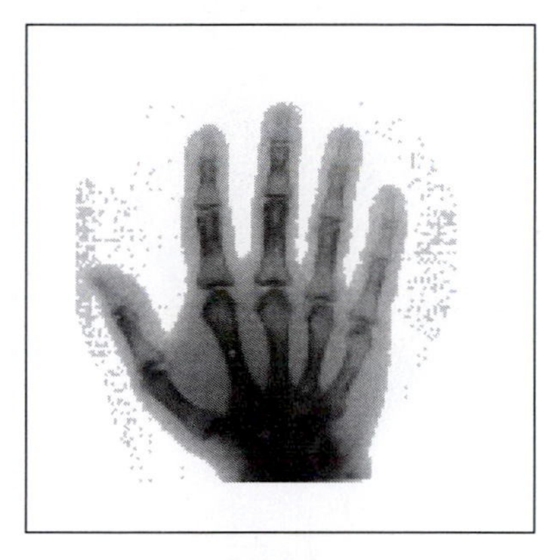
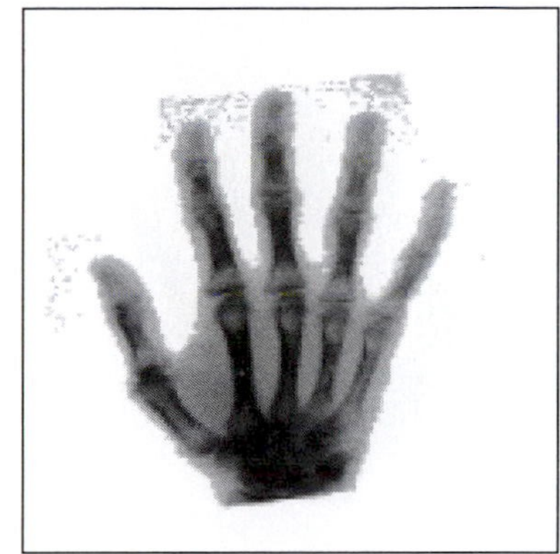
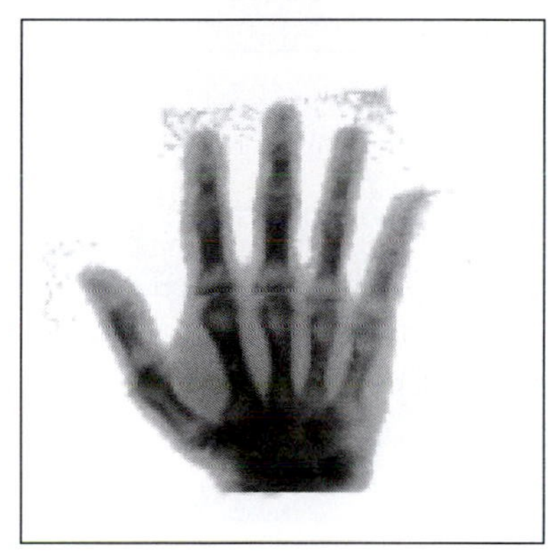
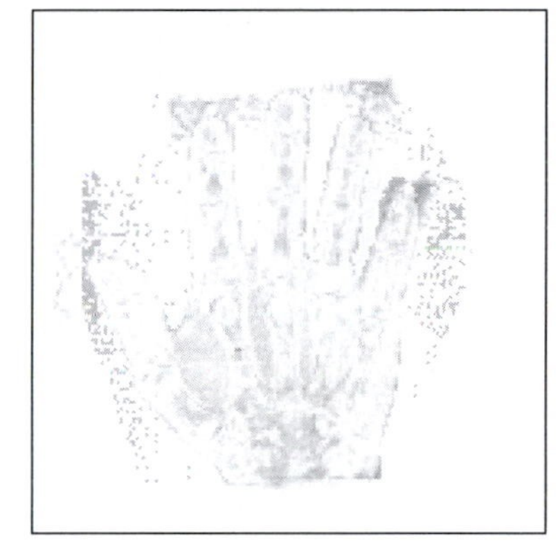

FIG. 11.6 Diffusion registration of hands using AOS, $\alpha = 5$, $\tau = 0.25$; TOP LEFT: reference image R, TOP RIGHT: template image T, BOTTOM LEFT: diffusion registered template T^{diff} with deformation, and BOTTOM RIGHT: difference $|R - T^{\text{diff}}|$.

11.6 Thirion's demons registration

In the landmark paper (1995), Thirion presents a new method to perform the non-rigid matching of two three-dimensional medical images. He describes his new method as based on so-called *demons*, a notation adapted from Maxwell's demons in thermodynamics.

The basic idea is to position demons at certain places in the image domain. These demons should then be able to decide whether or not a movement of one particle of the template image in a certain direction reduces the disparity between the reference and transformed images.

This intuitive idea is illustrated in Fig. 11.10. Roughly speaking, the task of the demons is to sort the particles. To do so they move the particles in- or outward depending on the relation between the scene and model. In order to

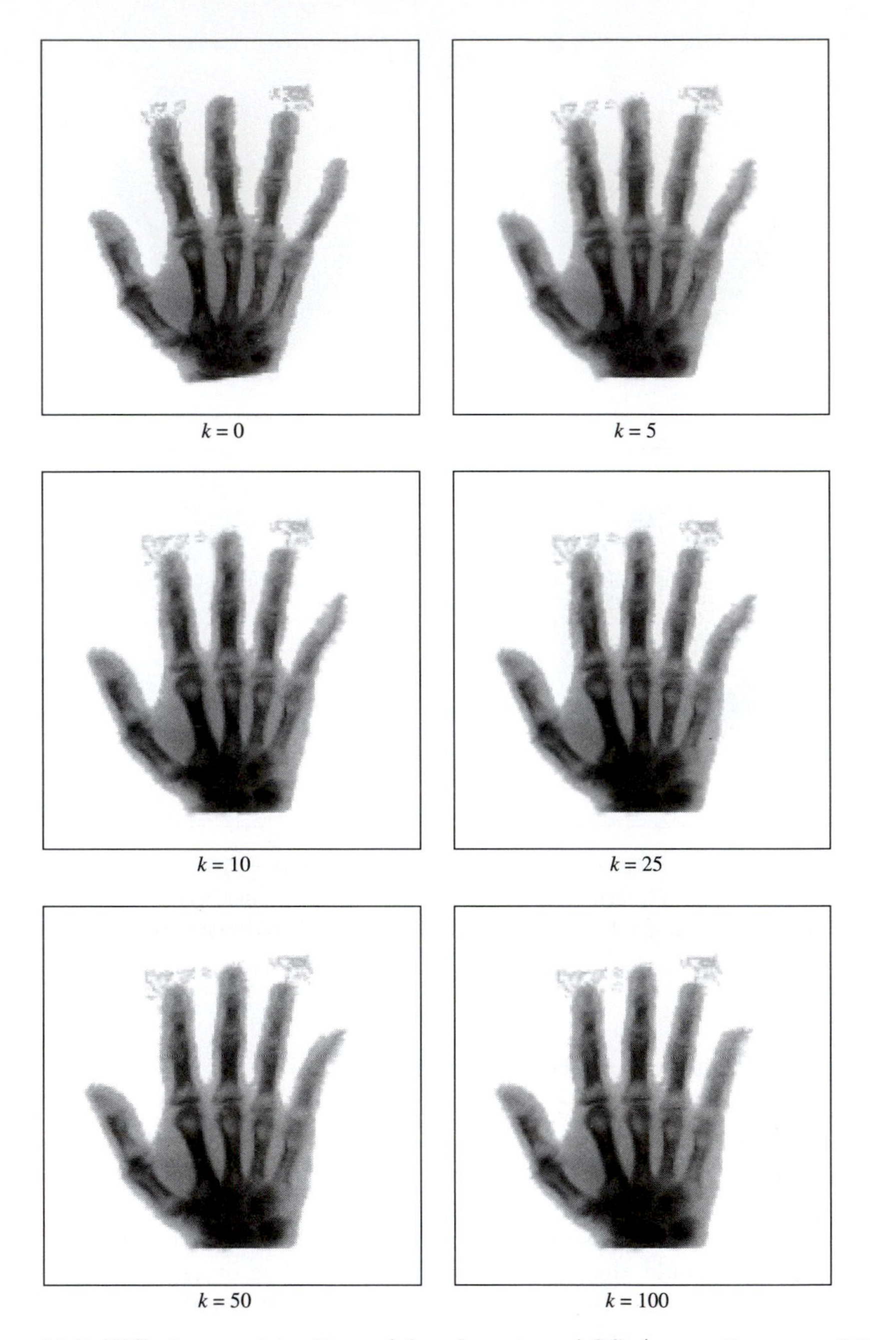

FIG. 11.7 Diffusion registration of hands using AOS ($\alpha = 5$, $\tau = 0.25$), intermediate results for $k = 0, 5, 10, 25, 50, 100$.

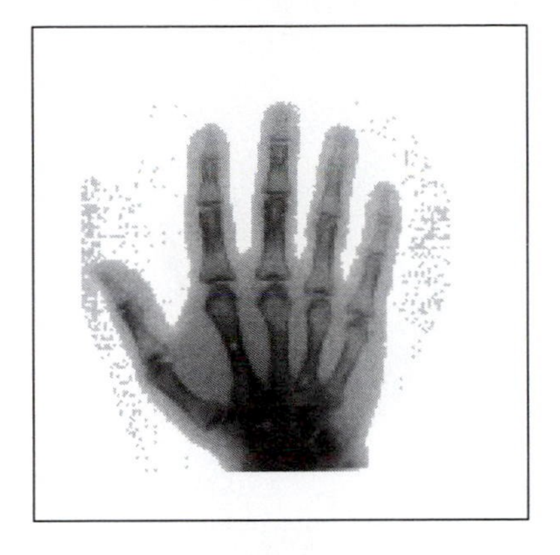
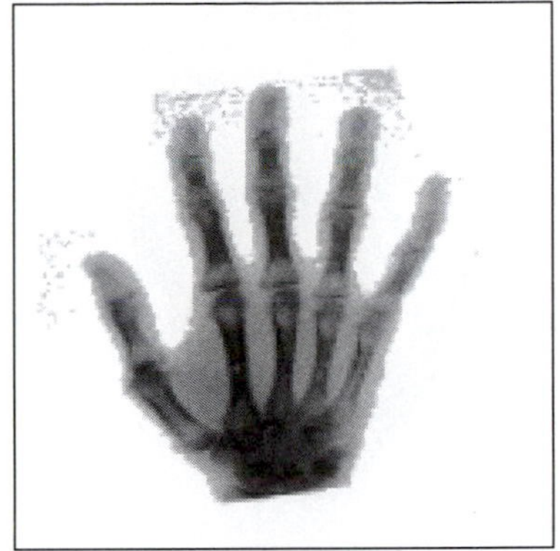
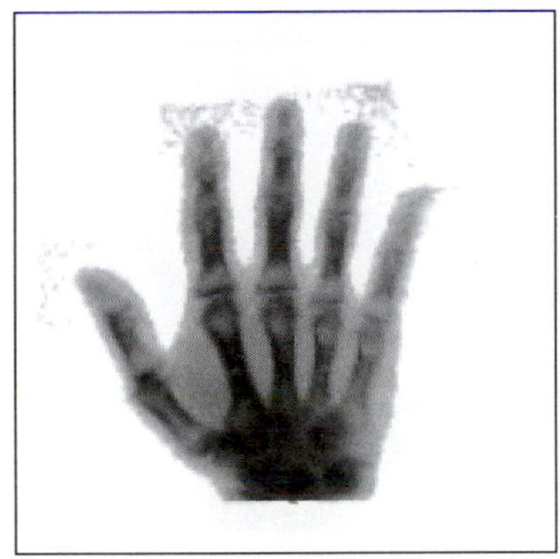
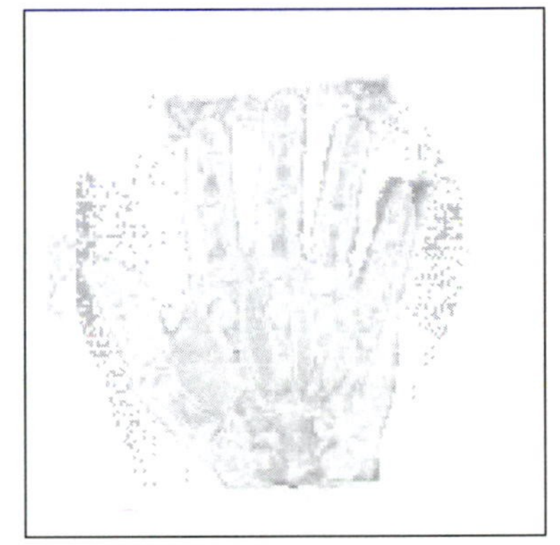

FIG. 11.8 Diffusion registration of hands using DCT ($\alpha = 5$, $\tau = 0.25$); TOP LEFT: reference image R, TOP RIGHT: template image T, BOTTOM LEFT: diffusion registered template T^{diff} with deformation, and BOTTOM RIGHT: difference $|R - T^{\text{diff}}|$.

illustrate this idea, two one-dimensional images R and T are considered, see Fig. 11.10. A demon d is located at a spatial position d, where $\nabla R(d) \neq 0$. Depending on the gradient $\nabla R(d)$ and the image difference $R(d) - T(d)$, the demons induce a pushing force p. The demon pushes the template according to $\nabla R(d)$, if $R(d) < T(d)$ (Fig. 11.10 middle), and according to $-\nabla R(d)$, if $R(d) > T(d)$ (Fig. 11.10 right).

A model algorithm is summarized in Algorithm 11.9, see also Thirion (1995). Thirion mentions many types of transformations. For the purpose of our considerations, we restrict ourselves to what Thirion called the *free-form case*. Following Thirion (1995), three questions remain open.

1. **Class of deformation**
   Demons are placed at each spatial position. The task then is to compute an appropriate displacement field from the typically rough force field. Here,

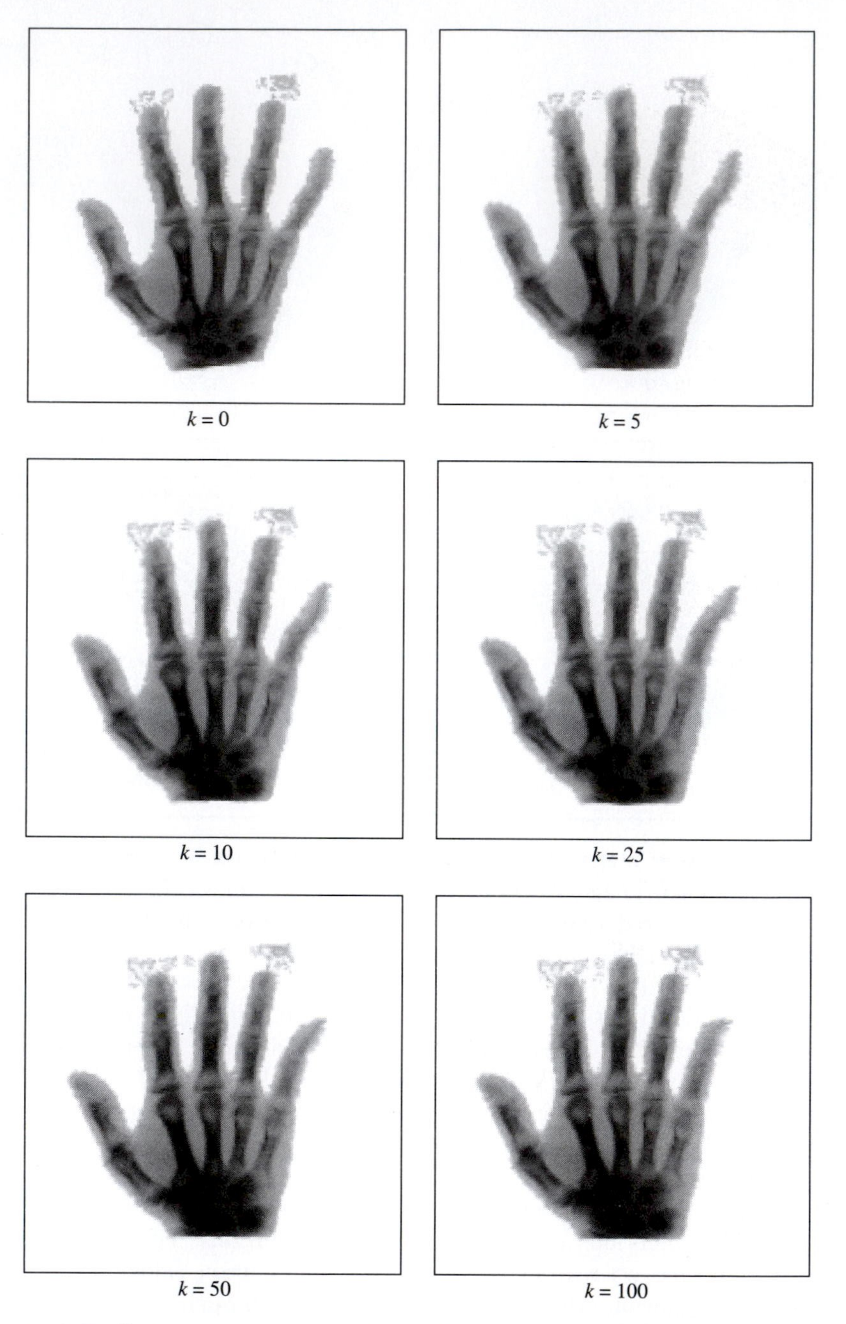

FIG. 11.9 Diffusion registration of hands with DCT ($\alpha = 5$, $\tau = 0.25$), intermediate results for $k = 0, 5, 10, 25, 50, 100$.

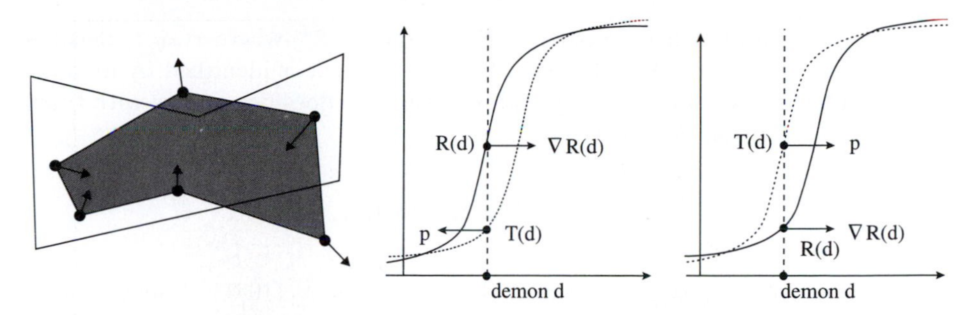

FIG. 11.10 Thirion's demons. LEFT: scene (gray) with six demons on its contour (big dots) and the contour of a deformable model; the arrows indicates movement, push direction is inward if the scene and model overlap and outward otherwise; MIDDLE and RIGHT: pushing and pulling force p at demon d in a one-dimensional model.

Algorithm 11.9 *Model algorithm for Thirion's demons registration.*

Let $k = 0$ and $\varphi^{(k)} = \mathcal{I}$ and D be a set of demons.
For $k = 0, 1, 2, \ldots$
  For each demon $d \in D$,
    compute the pushing force $f(d)$,
    according to the values of $\hat{T}(d) = T([\varphi^{(k)}]^{-1}(d))$ and $R(d)$.
  Compute a displacement field $u^{(k)} : \mathbb{R}^d \to \mathbb{R}^d$,
    based on $f(d)$, $d \in D$.
  Concatenate the transformations $\varphi^{(k)}$ and $u^{(k)}$,
    $\varphi^{(k+1)} = u^{(k)} \circ \varphi^{(k)}$.
End.

Thirion suggests a low-pass filter and in particular uses Gaussian filtering with a fixed and given covariance matrix σI_d.

2. **Interpolation scheme**
   An interpolation scheme has to be used. Here, we used a d-linear interpolation scheme; cf., Section 3.1.3.
3. **Type of demons**
   The demons type to be used is discussed in the next section.

11.6.1 *Demons type*

For simplicity we assume that the template T is a deformed version of the reference R, i.e., we assume a deformation process generating images $R(\cdot, t)$ for a time $t \in [0, 1]$, such that $R(x) = R(x, 0)$ and $T(x) = R(x, 1)$. Thus any particle P in the image domain Ω follows a path $x : \mathbb{R}^d \times [0,1] \to \mathbb{R}^d$, where $x(x_0, t)$ denotes the location of the particle at time t, where the particle is identified by its position x_0 at time zero. Since the gray scale of a particle does not change with time, we obtain the *optical flow equation*

$$0 = \frac{d}{dt} R(x,t) = \partial_t R(x,t) + (\nabla R(x,t))^\top \partial_t x,$$

where $\nabla = (\partial_{x_1}, \ldots, \partial_{x_d})^\top$, see also Horn & Schunck (1981). Using a finite difference approximation for the time derivative and setting $v := \partial_t x$, we obtain in particular

$$(\nabla R(x,t))^\top v \approx \frac{R(x,t+\tau) - R(x,t)}{\tau} \tag{11.21}$$

or

$$(\nabla R(x,t))^\top v \approx \frac{R(x,t) - R(x,t-\tau)}{\tau}, \tag{11.22}$$

where a forward and a backward finite difference approximation have been used. In particular for $\tau = 1$ we obtain

$$(\nabla R(x))^\top v = T(x) - R(x) \qquad (t = 0 \text{ in eqn (11.21)}) \tag{11.23}$$

and

$$(\nabla T(x))^\top v = T(x) - R(x) \qquad (t = 1 \text{ in eqn (11.22)}). \tag{11.24}$$

As in Thirion (1998) we focus on eqn (11.23) for the moment. Assuming that $\nabla R(x) \neq 0$, the general solution v^{general} is given by

$$v^{\text{general}} = \frac{T(x) - R(x)}{(\nabla R(x))^\top \nabla R(x)} \nabla R(x) + w,$$

where $w \in \operatorname{span}(\nabla R(x))^\perp$.

Thirion suggests selecting v^{shortest}, the general solution with smallest norm, i.e., $w = 0$. Note that the computation of v becomes delicate whenever $\|\nabla R(x)\|_{\mathbb{R}^d}$ is close to zero. Small perturbations may lead to large errors. To avoid this phenomenon, an additional regularization parameter $\kappa \neq 0$ is introduced, such that

$$v_\kappa^{\text{shortest}} := \frac{T(x) - R(x)}{(\nabla R(x))^\top \nabla R(x) + \kappa^2} \nabla R(x). \tag{11.25}$$

In order to reduce the number of parameters, Thirion suggests taking $\kappa = T(x) - R(x)$, ignoring the fact that $T(x) = R(x)$ occasionally. Note that, if $T(x) = R(x)$, the smallest norm solution is given by $v = 0$. However, the smallest norm solution is not well-defined by formula (11.25).

11.6.2 *Variational interpretation of Thirion's approach*

Now we relate Thirion's demon approach to diffusion registration. If we consider the particular distance measure

$$\mathcal{D}[u] = \frac{1}{2}\int_\Omega \frac{(R(x+u(x)) - T(x))^2}{\|\nabla R(x)\|^2_{\mathbb{R}^d} + \kappa^2} dx,$$

we find its Gâteaux derivative to be

$$\begin{aligned} d\mathcal{D}[u;v] &= \lim_{h\to 0}\frac{1}{2}\int_\Omega \frac{(R(x+u+hv) - T(x))^2 - (R(x+u)-T(x))^2}{\|\nabla R(x)\|^2_{\mathbb{R}^d} + \kappa^2} dx \\ &= \int_\Omega \frac{(R(x+u) - T(x))}{\|\nabla R(x)\|^2_{\mathbb{R}^d} + \kappa^2} \nabla R(x)^\top v \, dx. \end{aligned}$$

Thus, Thirion's velocity $v_\kappa^{\text{shortest}}$ might be viewed as a force f in the variational setting.

If in addition the smoother $\mathcal{S}^{\text{diff}}$ introduced in eqn (11.1) is used to privilege smooth displacements, the Euler–Lagrange equations are given by Theorem 11.1.

A numerical scheme may be based on the semi-implicit scheme (8.8). If we had to solve (8.8) with respect to the whole space, i.e., $\Omega = \mathbb{R}^d$, then, under mild conditions on the driving force $f^{(k)}(x) := f(x, u(x, t_k))$, it would be possible to come up with an analytic solution. A representative result in this direction reads (see Folland (1995, Th. 4.8)): if $f^{(k)} \in L_1$, then the convolution

$$\begin{aligned} u^{(k+1)}(x,t) &= K_t(x) * f^{(k)}(x) \\ &= \int_{-\infty}^{t}\int_{\mathbb{R}^d} K_{t-s}(x-y) f^{(k)}(y) dy \, ds, \quad t > 0, \end{aligned}$$

is well-defined almost everywhere and is a distributional solution of (8.8). It will even be a classical solution if $f^{(k)} \in C^p$ for $p > 1$. Here

$$K_t(x) = (4\pi t)^{-d/2} \exp\left(- \|x\|^2_{\mathbb{R}^d} / (4t)\right)$$

denotes the Gauss kernel. In order to solve (8.8) with respect to the discretized bounded region $\Omega =]0,1[^d$, one may approximate the Gauss kernel by a Gauss filter of suitable length; that is, to compute at each time-step the force convolved with a Gauss filter K_σ of characteristic width σ. As is well-known, the Gauss filter-based scheme is less accurate than the outlined finite difference scheme.

This Gauss filter-based approach is essentially what Thirion calls "Demons 1: a complete grid of demons"; see Thirion (1998). However, he gives no hint on how to choose the parameter σ for a given application. It turns out in practice that a proper choice of this free parameter is a delicate matter. On the other hand, independent of the choice of σ, Thirion's approach is also of linear complexity.

11.6.3 *Steepest descent interpretation*

In this section we give another straightforward interpretation of Thirion's approach. The overall goal is to minimize the distance measure $\mathcal{D}[R,T;u]$; cf., eqn (8.2). Starting with a smooth displacement u one might use a steepest descent method $\partial_t u = -f(\cdot,u)$, see Theorem 8.1, or in a time discrete setting,

$$u^{(k+1)} = u^{(k)} + \tau f(\cdot, u^{(k)}).$$

The problem is that f is in general not smooth, such that the update and hence $u^{(k+1)}$ might be non-smooth. The trick is to project the update onto a smooth space, e.g., by convolving with a Gauss kernel. This gives

$$u^{(k+1)} = K_\sigma * (u^{(k)} + \tau f(\cdot, u^{(k)})).$$

11.7 Discussion of diffusion registration

The main drawback of diffusion registration is that although it is obvious to measure smoothness by oscillations of the gradients, it is not physical. Even though each component u_ℓ of the displacement can be viewed as a solution of a particular heat equation, a physical interpretation for the vector field u is missing. In our experience, however, the non-physical behavior of the method can hardly be detected in "real-life" applications.

In the registration process, the spatial directions are coupled only through the forces. It is this property that can be viewed as one major advantage of diffusion registration. The spatial decoupling allows for a block diagonalization. In addition, for each block the AOS scheme (cf., Section 11.3) presents a fast and stable solution technique of linear complexity. This makes diffusion registration a very attractive registration scheme, in particular for high-dimensional image data.

12

CURVATURE REGISTRATION

Elastic, fluid, and diffusion registration are sensitive with respect to affine linear displacements; see, e.g., Section 10.6. In particular, for dimension $d = 2$ and

$$u(x) = Cx + b, \quad \text{with} \quad C \in \mathbb{R}^{2\times 2}, \quad b \in \mathbb{R}^2,$$

for the elastic potential of u, we have

$$\mathcal{P}[u] = \mathcal{S}^{\mathrm{elas}}[u] = \int_\Omega \mu(c_{1,1}^2 + (c_{1,2} + c_{2,1})^2/2 + c_{2,2}^2) + \lambda(c_{1,1} + c_{2,2})^2/2\, dx.$$

Hence, for $\mu \neq 0$,

$$\mathcal{P}[u] = 0 \iff c_{1,1} = c_{2,2} = 0 \wedge c_{1,2} = -c_{2,1}.$$

Note that in the derivation of elastic registration (see Section 9.1), we explicitly decomposed the transformation into rigid and non-rigid parts. Thus, from this modeling, the rigid parts need to be pre-registered.

For diffusion registration, we have

$$\mathcal{S}^{\mathrm{diff}}[u] = \int_\Omega c_{1,1}^2 + c_{1,2}^2 + c_{2,1}^2 + c_{2,1}^2 dx$$

and $\mathcal{S}^{\mathrm{diff}}[u] = 0 \iff C = 0$. Since diffusion registration penalizes the norm of the gradient, this property is a direct consequence of the regularizing term.

As a consequence, for all these non-linear registration techniques an affine linear pre-registration is unavoidable. In order to circumvent this additional pre-registration we introduce a novel regularizing term based on second order derivatives. Since the regularizer is related to *curvature*, the novel registration is called *curvature registration*; cf., Fischer & Modersitzki (2003).

The main point is not that the additional pre-registration becomes redundant but that the registration becomes less dependent on the initial position of the reference and template images.

Curvature registration is based on the distance measure $\mathcal{D}$ (cf., eqn (8.10)) and the regularizer

$$\mathcal{S}^{\mathrm{curv}}[u] := \tfrac{1}{2} a[u,u], \tag{12.1}$$

where the bi-linear form a is defined by

$$a[u,v] = \sum_{\ell=1}^{d} \int_\Omega \Delta u_\ell \Delta v_\ell \, dx$$

and Neumann boundary conditions $\nabla u_\ell = \nabla \Delta u_\ell = 0$ for $x \in \partial\Omega$, $\ell = 1, \dots, d$, are imposed.

The integrand $(\Delta u_\ell)^2$ of $\mathcal{S}^{\text{curv}}$ might be viewed as an approximation to the curvature. Thus, the idea of the regularizer is to minimize the curvature of the components of the displacement.

12.1 Continuous and discrete bi-harmonic equations

The Euler–Lagrange equations for the joint functional of Problem 8.1 are summarized in Theorem 12.1.

Theorem 12.1 *The* Euler–Lagrange *equations for* $\mathcal{J}^{\text{curv}} = \mathcal{D}^{\text{SSD}} + \alpha\mathcal{S}^{\text{curv}}$, *where* $\mathcal{D}^{\text{SSD}}$ *is defined by eqn (8.10) and* $\mathcal{S}^{\text{SSD}}$ *is defined by eqn (12.1) are*

$$f(x, u(x)) + \alpha\Delta^2 u(x) = 0, \quad x \in \Omega, \tag{12.2}$$

$\nabla u_\ell = \nabla\Delta u_\ell = 0$ *for* $x \in \partial\Omega$, $\ell = 1, \dots, d$.

Proof Follows from Theorem 8.1 and, with $\widetilde{a}[\xi, \eta] = \int_\Omega \Delta\xi\Delta\eta\, dx$, from

$$\begin{aligned}
\widetilde{a}[\xi, \eta] &= \int_\Omega \Delta\xi\Delta\eta\, dx \\
&= \int_{\partial\Omega} \Delta\xi \,\langle \nabla\eta, n\rangle_{\mathbb{R}^d}\, dx - \int_\Omega \langle \nabla\Delta\xi, \nabla\eta\rangle_{\mathbb{R}^d}\, dx \\
&= -\int_{\partial\Omega} \eta \,\langle \nabla\Delta\xi, n\rangle_{\mathbb{R}^d}\, dx + \int_\Omega \eta\Delta^2\xi\, dx \\
&= \int_\Omega \eta\Delta^2\xi\, dx.
\end{aligned}$$

Here, Green's formula and the Neumann boundary conditions have been utilized. □

As an important fact, we note that the partial differential operator related to the smoother $\mathcal{S}^{\text{curv}}$ is just the second power of the partial differential operator arising in diffusion registration, cf., Theorem 11.1. As a consequence, we employ a numerical scheme for the discrete Euler–Lagrange equations

$$f(\vec{X}, \vec{U}) + \alpha I_d \otimes A^{\text{curv},d}\vec{U} = 0, \tag{12.3}$$

where $A^{\text{curv},d} = (A^{\text{diff},d})^\top A^{\text{diff},d} = (A^{\text{diff},d})^2$; see also eqn (11.4).

The Euler–Lagrange equations for the curvature registration functional is also known as the *bi-harmonic* equation; see, e.g., Hackbusch (1987, §5.3).

As a by-product of the analysis presented in Section 11.2 we have the following corollary; see also Theorem 11.3.

Corollary 12.2 *Let S be a d-dimensional, symmetric matrix stencil and $A^{(d)}$ be the matrix representation of the convolution with $S * S$ with respect to Neumann boundary conditions. Then*

$$\begin{aligned}(D^{(d)})^2 &:= (V_{n_d} \otimes \cdots \otimes V_{n_1})^\top \; A^{(\mathrm{curv},d)} \; (V_{n_d} \otimes \cdots \otimes V_{n_1}) \\ &= \mathrm{diag}((d_{j_1,\ldots,j_d})^2, \; j_q = 1,\ldots,n_q, \; q = 1,\ldots,d)\end{aligned}$$

where V_m is defined by eqn (11.10) and $d_{j_1,\ldots,j_d}$ is defined recursively by

$$\begin{aligned} d_{j_1}^{p_{d-1},\ldots,p_1} &= S_{2,p_{d-1},\ldots,p_1} + 2S_{2,p_{d-1},\ldots,p_1} \cos \tfrac{j_1\pi}{n_1}, \\ d_{j_1,\ldots,j_k}^{p_{d-k},\ldots,p_1} &= d_{j_1,\ldots,j_k}^{2,p_{d-k},\ldots,p_1} + 2d_{j_1,\ldots,j_k}^{1,p_{d-k},\ldots,p_1} \cos \tfrac{j_k\pi}{n_k}, \\ d_{j_1,\ldots,j_d} &= d_{j_1,\ldots,j_{d-1}}^2 + 2d_{j_1,\ldots,j_{d-1}}^1 \cos \tfrac{j_d\pi}{n_d}. \end{aligned}$$

The matrix stencils $S^{\mathrm{curv},d} = S^{\mathrm{diff},d} * S^{\mathrm{diff},d}$ for dimension $d = 2, 3$ are summarize in Table 12.1; for $S^{\mathrm{diff},d}$ see Table 11.1.

As a further by-product of the analysis in Section 11.2, we note that an $\mathcal{O}(n \log n)$ implementation based on the discrete cosine transformation can be deduced. Here n denotes the number of voxels.

Table 12.1 *Matrix stencils $S^{\mathrm{curv},d} = S^{\mathrm{diff},d} * S^{\mathrm{diff},d}$ for the bi-harmonic operator and dimension $d = 2, 3$; $S^{\mathrm{diff},d}$ are given in Table 11.1.*

$$S^{\mathrm{curv},2} = \begin{pmatrix} 0 & 0 & 1 & 0 & 0 \\ 0 & 2 & -8 & 2 & 0 \\ 1 & -8 & 20 & -8 & 1 \\ 0 & 2 & -8 & 2 & 0 \\ 0 & 0 & 1 & 0 & 0 \end{pmatrix},$$

$$\begin{aligned} S^{\mathrm{curv},3} &= \left(S^{\mathrm{curv},3}_{j,k,\ell}\right)_{j,k,\ell=1,\ldots,5} \\ &= \begin{cases} 42, & j=k=\ell=2, \\ -12, & |j-3|+|k-3|+|\ell-3|=1, \\ 2, & |j-3|+|k-3|+|\ell-3|=2 \\ & \wedge \max\{|j-3|+|k-3|+|\ell-3|\}=1, \\ 1, & |j-3|+|k-3|+|\ell-3|=2 \\ & \wedge \max\{|j-3|+|k-3|+|\ell-3|\}=2, \\ 0, & \text{otherwise.} \end{cases} \end{aligned}$$

In contrast to diffusion registration, where even an $\mathcal{O}(n)$ algorithm can be obtained by exploiting an additive operator splitting scheme for a time marching-based implementation, see Section 11.3, curvature registration does not permit such a fast numerical treatment.

This can already be seen for dimension $d = 2$, where the discrete Euler–Lagrange equations (cf., eqn (12.3)) read

$$\frac{\vec{V}^{(k+1)} - \vec{V}^{(k)}}{\tau} + (A_1 + A_2)^2 \vec{V}^{(k+1)} = \vec{F}_j^{(k)},$$

where $\vec{V}^{(k)}$ and $\vec{F}_j^{(k)}$ are defined in eqns (11.15) and (11.16), respectively. The matrices $A_1, A_2 \in \mathbb{R}^{n \times n}$ are given by

$$A_1 = \alpha\, I_{n_2} \otimes M_{n_1}, \quad A_2 = \alpha\, M_{n_2} \otimes I_{n_1},$$

where $n := n_1 n_2$ is the number of pixels, and M_m is defined in eqn (11.5). Using these particular matrices, an AOS scheme for curvature registration reads

$$\vec{V}^{(k+1)} = \frac{1}{3} \sum_{\ell=1}^{3} \vec{V}_\ell^{(k+1)},$$

where

$$\left(I_n + 3\tau I_{n_2} \otimes M_{n_1}^2\right) \quad \vec{V}_1^{(k+1)} = \vec{V}^{(k)} + \tau \vec{F}_j^{(k)}, \tag{12.4}$$

$$\left(I_n + 6\tau M_{n_2} \otimes M_{n_1}\right) \; \vec{V}_2^{(k+1)} = \vec{V}^{(k)} + \tau \vec{F}_j^{(k)}, \tag{12.5}$$

$$\left(I_n + 3\tau M_{n_2}^2 \otimes I_{n_1}\right) \quad \vec{V}_3^{(k+1)} = \vec{V}^{(k)} + \tau \vec{F}_j^{(k)}. \tag{12.6}$$

For the solution of eqns (12.4) and (12.6), an $\mathcal{O}(n)$ algorithm is available. However, the overall complexity is governed by the solution of eqn (12.5), where only $\mathcal{O}(n \log n)$ direct solution schemes are known. Thus, our implementation is directly based on the factorization presented in Corollary 12.2.

12.2 A relation to thin plate splines

In Chapter 4, the bi-linear form

$$a_q^{\mathrm{TPS}}[\xi, \eta] = \sum_{|\kappa|=q} c_\kappa \int_\Omega D^\kappa \xi\, D^\kappa \eta\; dx$$

was investigated; see, Section 4.3.1. In fact, there is a strong relation between this spline approach and our curvature registration. In the spline approach, one is looking for a u_ℓ which interpolates or approximates certain data but is as smooth as possible. Smoothness is measured in terms of $a_q^{\mathrm{TPS}}[u_\ell, u_\ell]$, which might also be viewed as a curvature.

From the proof of Theorem 12.1, we have

$$\begin{aligned} a^{\text{curv}}[\xi,\eta] &= \int_\Omega \Delta\xi\Delta\eta\,dx \\ &= \int_{\partial\Omega} \Delta\xi\,\langle\nabla\eta,n\rangle_{\mathbb{R}^d} - \eta\,\langle\nabla\Delta\xi,n\rangle_{\mathbb{R}^d}\,dx + \int_\Omega \eta\Delta^2\xi\,dx. \end{aligned}$$

In particular for $q=2$, we find

$$\begin{aligned} a_2^{\text{TPS}}[\xi,\eta] &= \int_\Omega \partial_{x_1x_1}\xi\partial_{x_1x_1}\eta + 2\partial_{x_1x_2}\xi\partial_{x_1x_2}\eta + \partial_{x_2x_2}\xi\partial_{x_2x_2}\eta\,dx \\ &= \int_\Omega \langle\nabla\partial_{x_1}\xi,\nabla\partial_{x_1}\eta\rangle_{\mathbb{R}^d} + \langle\nabla\partial_{x_2}\xi,\nabla\partial_{x_2}\eta\rangle_{\mathbb{R}^d}\,dx \\ &= \int_{\partial\Omega} \partial_{x_1}\eta\,\langle\nabla\partial_{x_1}\xi,\vec{n}\rangle_{\mathbb{R}^d} + \partial_{x_2}\eta\,\langle\nabla\partial_{x_2}\xi,\vec{n}\rangle_{\mathbb{R}^d}\,dx \\ &\quad - \int_\Omega \partial_{x_1}\eta\Delta[\partial_{x_1}\xi] + \partial_{x_2}\eta\Delta[\partial_{x_2}\xi]\,dx \\ &= \int_{\partial\Omega} \nabla\eta[\nabla^2\xi]\vec{n}\,dx - \int_\Omega \langle\nabla\Delta\xi,\nabla\eta\rangle\,dx \\ &= \int_{\partial\Omega} \nabla\eta[\nabla^2\xi]\vec{n} - \eta\,\langle\nabla\Delta\xi,\vec{n}\rangle_{\mathbb{R}^d}\,dx + \int_\Omega \eta\Delta^2\xi\,dx. \end{aligned}$$

It is worthwhile noticing that the derivatives of the two functionals $a^{\text{curv}}[\xi,\xi]$ and $a_2^{\text{TPS}}[\xi,\xi]$ share the same main part.

12.3 Curvature registration: squares

We begin with a small academic example, given the reference and template images depicted in Fig. 12.1. We perform both curvature and fluid registration. In particular we choose $\alpha = 10^6$, $\tau = 5$, $\mu = 10^3$, and $\lambda = 0$. The deformed grids are also illustrated in Fig. 12.1. Intermediate results are depicted in Fig. 12.2.

Both curvature and fluid registration give meaningful registration results. For both techniques, the deformed template images are (besides some interpolation artifacts) identical to the reference image, and the deformation is visually pleasing. However, the deformations are completely different. For curvature registration we detect an almost linear deformation of the original image. Note that affine linear rergistrations are not penalized by the curvature regularizer. On the other hand, for fluid registration, the deformation is not linear at all.

This example also illustrates that the quality of registration cannot be measured by inspecting the deformed template or the difference image. Moreover, for this particular example, it is not possible to decide which of the two registrations

FIG. 12.1 TOP LEFT: reference image, TOP RIGHT: template image, BOTTOM LEFT: deformed grid after curvature registration ($\alpha = 10^6$ and $\tau = 5$), and BOTTOM RIGHT: deformed grid after fluid registration ($\mu = 10^3$ and $\lambda = 0$).

is better. The problem is that image registration is an ill-posed problem. The distance measure can have two different meaningful solutions, the regularizer term deciding which one is more likely. One topic of regularization is to distinguish a likely solution on purpose. Note that this ambiguity is intrinsic to the problem and has nothing to do with "complicated" regularizing terms. Even when restricted to the class of rigid transformations, the actual example has multiple solutions, e.g., the particular translation and a rotation of about 180 degrees.

12.4 Curvature registration: • to C

We continue the registration of a • to a C, see also Section 9.8, Section 10.5, and Section 11.4. In Fig. 12.3, we show the intermediate results of curvature registration for $k = 0(20)100$. In order to visualize the deformation, the template

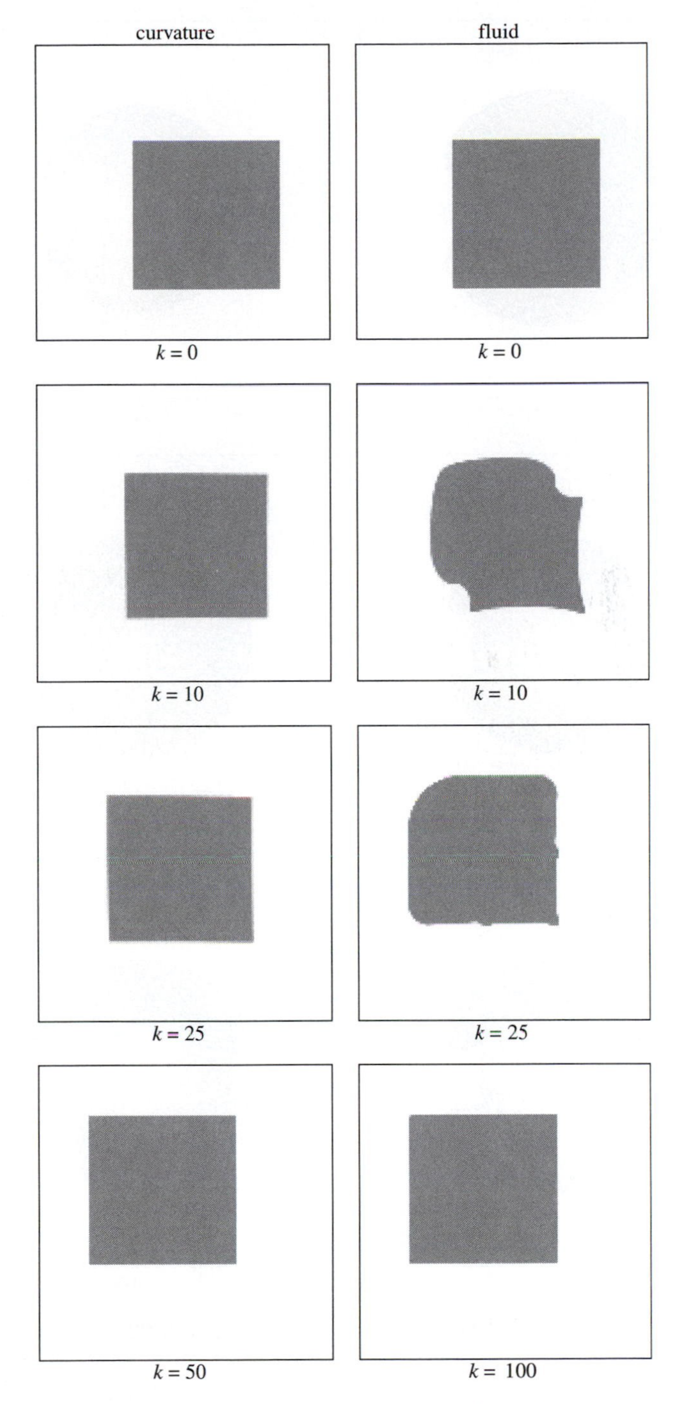

FIG. 12.2 Curvature (LEFT) and fluid (RIGHT) registration of squares: intermediate results.

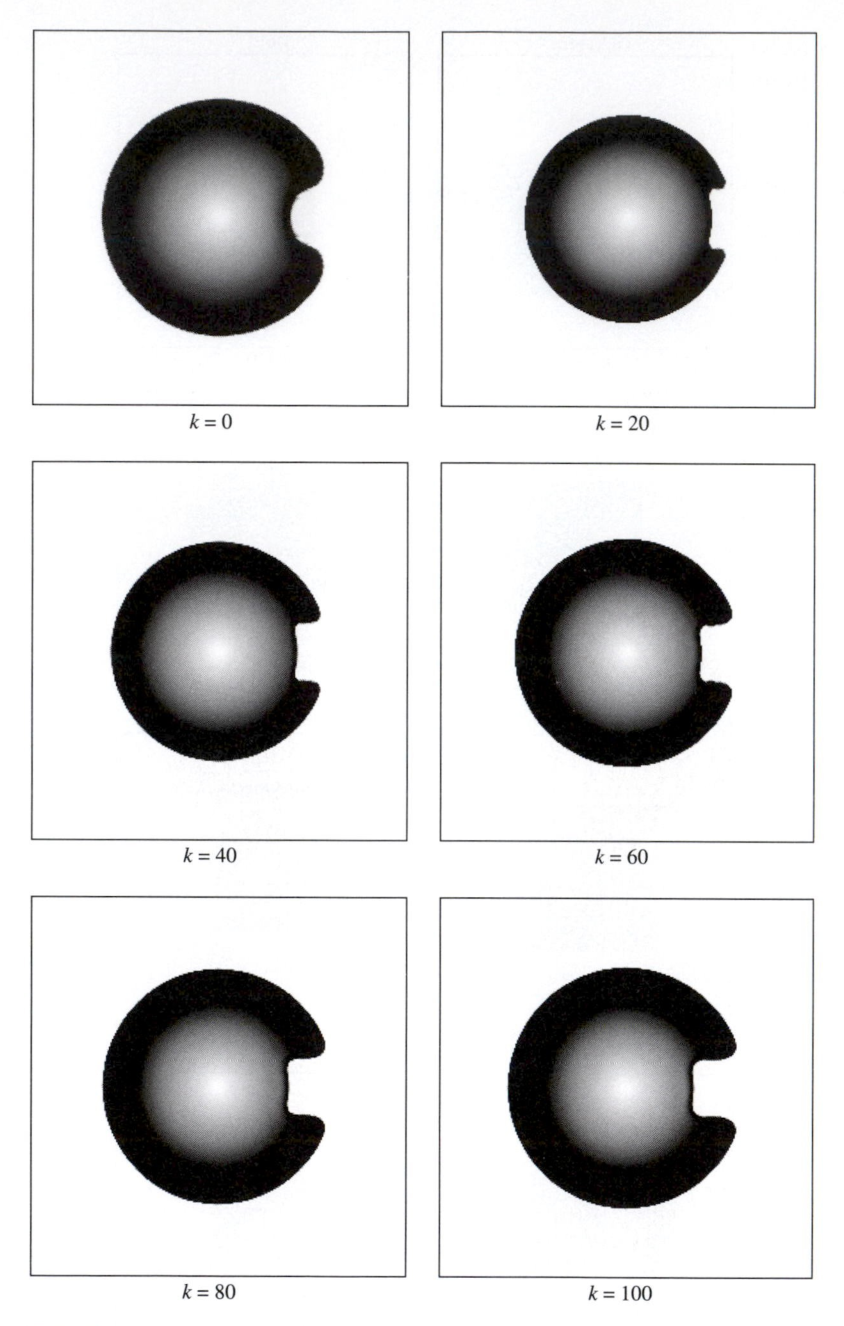

FIG. 12.3 Curvature registration of • to C, $\alpha = 50$, $\tau = 0.02$; intermediate results for $k = 0(20)100$.

has been shaded. Note that the computations have been performed on the original images. In this example, we chose $\alpha = 50$ and $\tau = 0.02$.

12.5 Curvature registration: hands

Figure 12.4 shows the two modified X-ray images of human hands and the diffusion registered hand with an illustration of the deformation, see also Section 9.9, Section 10.6, and Section 11.5. In Fig. 12.5 we show the intermediate results for various iterations.

Finally, in Figs 12.6 and 12.7 we show the results of the registration performed on the images but without performing an affine linear pre-registration step. Though the number of iterations needed to obtain this result is quite high, this example clearly indicates that curvature registration gives much better

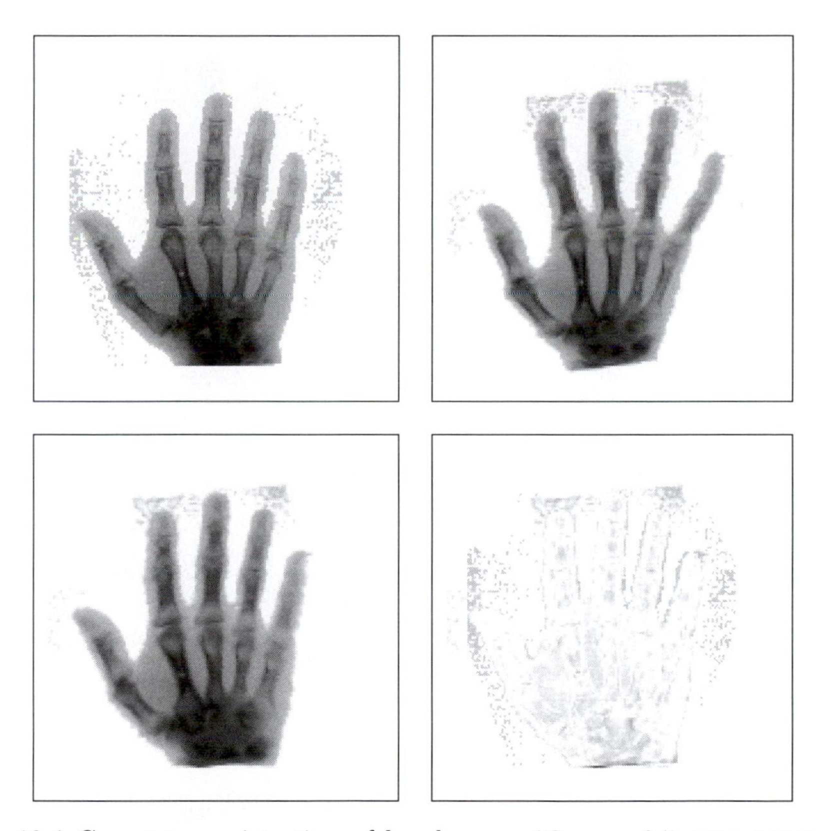

FIG. 12.4 Curvature registration of hands, $\alpha = 50$, $\tau = 0.5$; TOP LEFT: reference image R, TOP RIGHT: template image T, BOTTOM LEFT: curvature registered template T^{curv} with deformation, and BOTTOM RIGHT: difference $|R - T^{\text{curv}}|$.

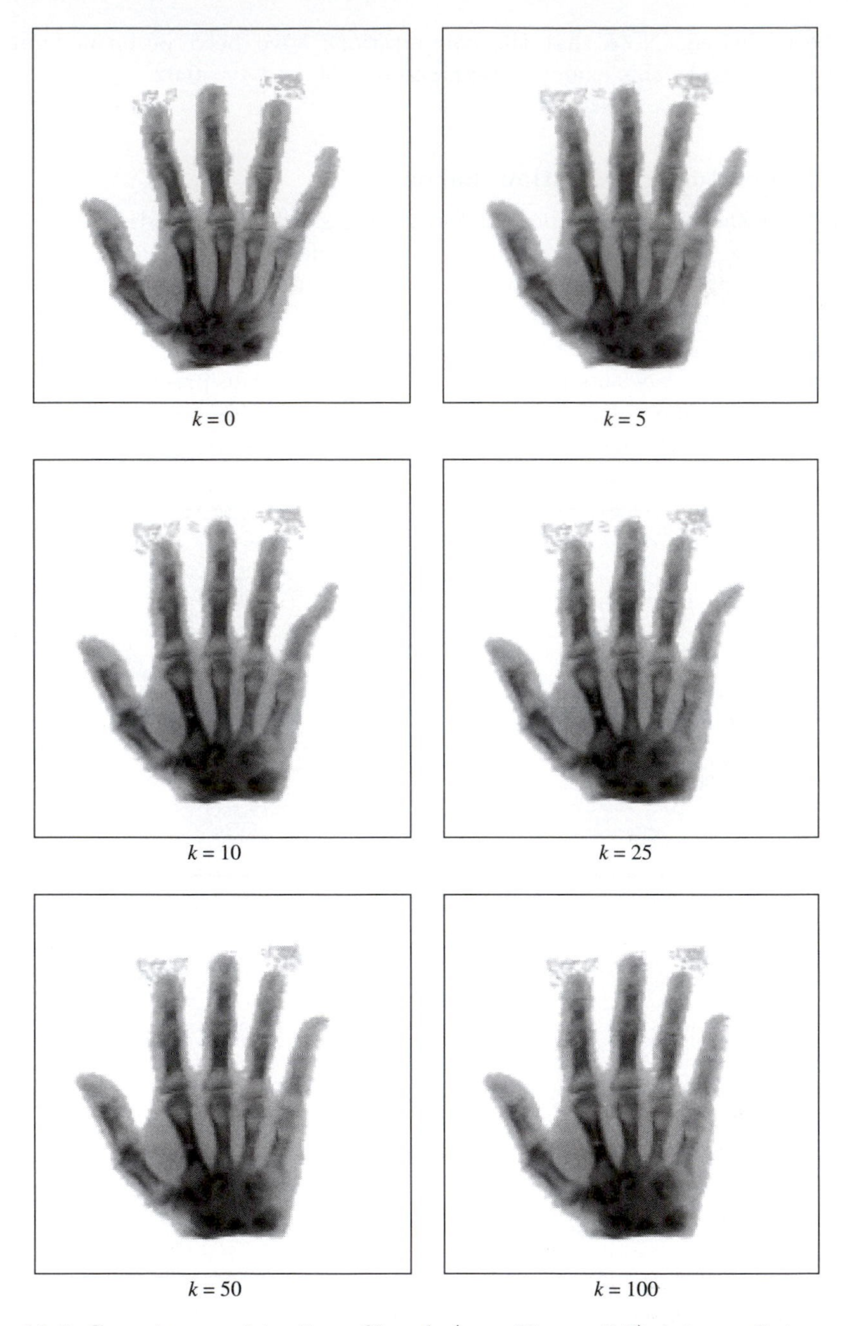

FIG. 12.5 Curvature registration of hands ($\alpha = 50$, $\tau = 0.5$), intermediate results for $k = 0, 5, 10, 25, 50, 100$.

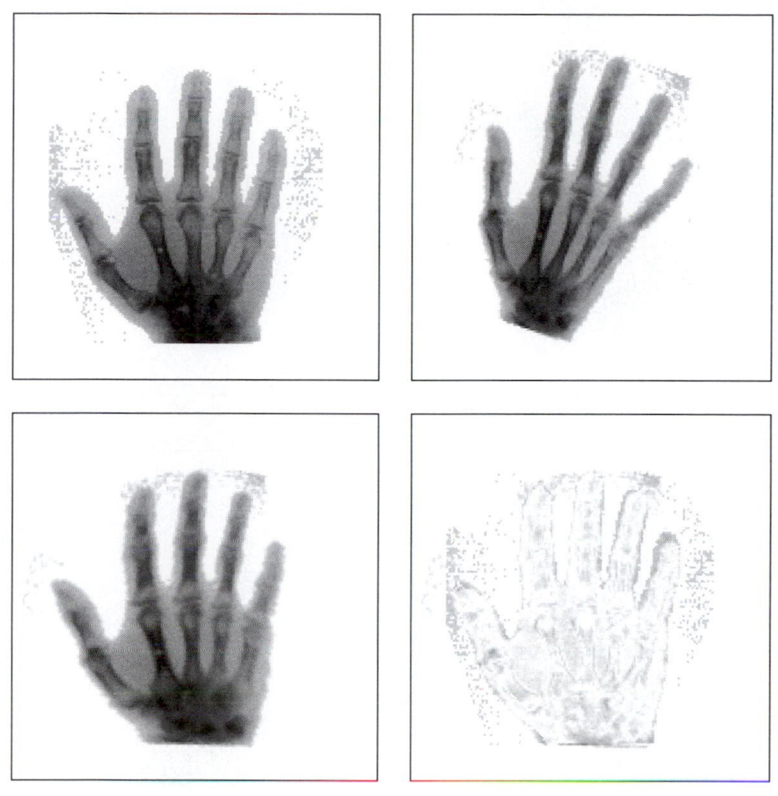

FIG. 12.6 Curvature registration of hands without pre-registration, $\alpha = 10^5$, $\tau = 1$; TOP LEFT: reference image R, TOP RIGHT: template image T, BOTTOM LEFT: curvature registered template T^{curv} with deformation, and BOTTOM RIGHT: difference $|R - T^{\text{curv}}|$.

registration results than, for example, fluid registration; cf., Section 10.6 and in particular Figs 10.5 and 10.6.

12.6 Discussion of curvature registration

Curvature registration has been designed for particular applications, for which a reliable, affine linear pre-registration cannot be guaranteed. Elastic, fluid, and diffusion registration depend on the pre-registration step; cf., Section 10.6. Since affine linear mappings belong to the kernel of the regularizer $\mathcal{S}^{\text{curv}}$, curvature registration performs better for these applications; cf., Section 12.3. However, the initial position plays an important role also for curvature registration and it is not advisable to skip the pre-registration.

The curvature registration regularizer is based on second order derivatives. Thus, the transformation is smoother than the ones obtained by a first order derivatives regularizer, like, e.g., diffusion registration.

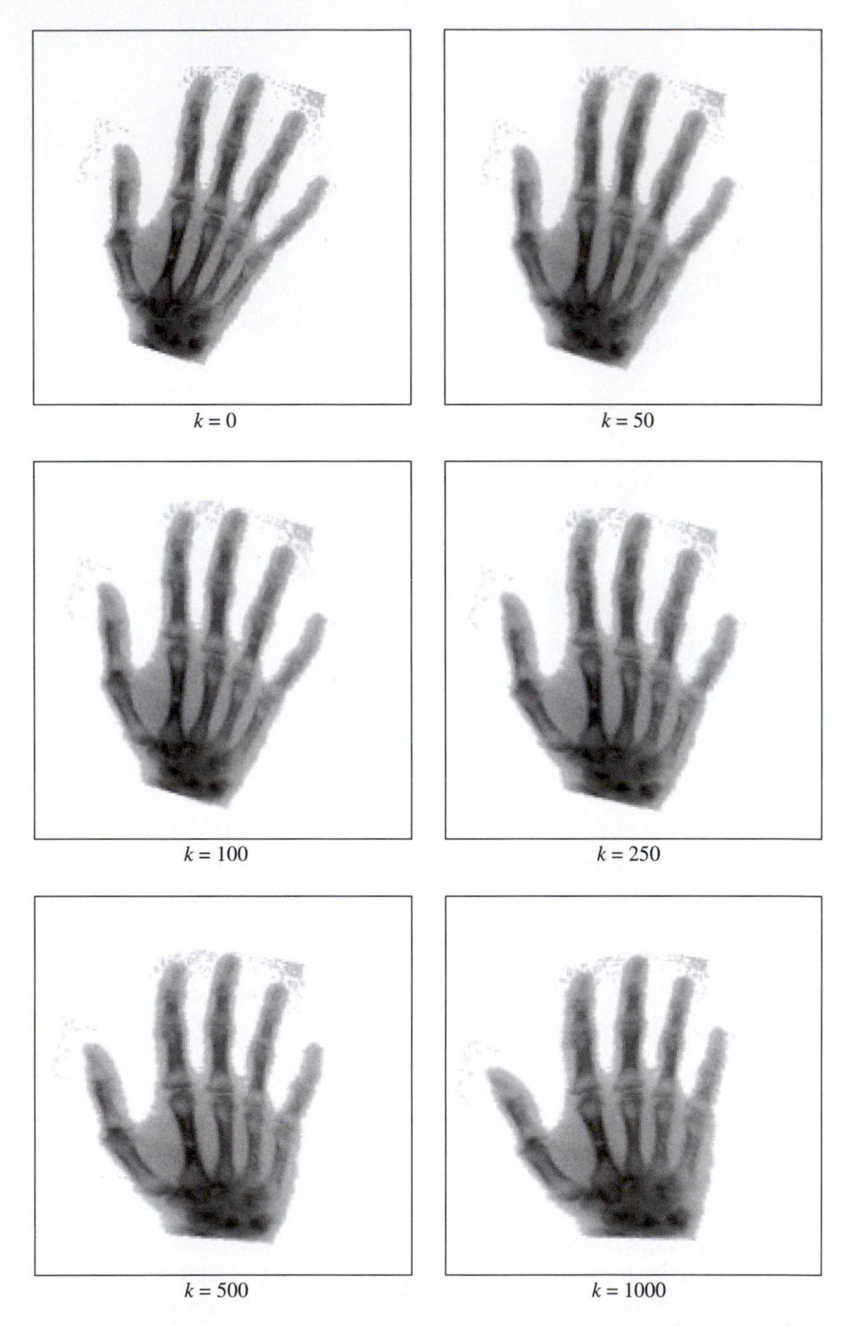

FIG. 12.7 Curvature registration of hands without pre-registration ($\alpha = 50$, $\tau = 0.5$), intermediate results for $k = 0, 50, 100, 250, 500, 1000$.

13

CONCLUDING REMARKS

The purpose of this chapter is to briefly summarize the proposed non-parametric registration approach as well as the presented schemes and to briefly discuss some of the open questions. The unified approach to non-parametric image registration is presented in Chapter 8. The basic ingredients are a distance measure $\mathcal{D}$ and a regularizer $\mathcal{S}$. The distance measure is the driving force of the registration and the regularizer is used to restrict the transformation to an appropriate class. The required displacement u is a solution of

$$\mathcal{D}[R, T; u] + \alpha \mathcal{S}[u] = \min,$$

with some additional boundary conditions. Here, we used

$$\mathcal{D}[R, T; u] := \mathcal{D}[R, T; u] := \tfrac{1}{2} \left\| T_u - R \right\|_{L_2(\Omega)};$$

cf., Definition 6.1 (see also Theorem 8.1). For the regularizer $\mathcal{S}$, we discussed four particular choices which then lead to the elastic, fluid, diffusion, and curvature registration schemes.

The numerical treatment presented is based on the corresponding Euler–Lagrange equations, which lead to a system of non-linear partial differential equations, $\mathcal{A}[u] = f(\,\cdot\,, u)$; cf., eqn (8.5). A fixed-point or time marching algorithm is used in order to circumvent the non-linearity of f with respect to u. After appropriate finite difference approximations of the time-discrete equations, we ended up with an iterative scheme, where in each step a large system of linear equation has to be solved. Particular solution techniques are designed to reduce the numerical complexity of the overall iterative schemes.

In Section 13.1, we briefly summarize the elastic, fluid, diffusion, and curvature registration schemes. In Section 13.2, we comment on timings obtained in a particular environment. A remark on how to compare the different approaches is given in Section 13.4. We also comment on the choice of the parameters λ, μ and α, τ, respectively (cf., Section 13.5), on further acceleration techniques of the schemes (cf., Section 13.6), and on extensions (cf., Section 13.7).

13.1 Summary of non-parametric image registration

Elastic registration

- Regularization is based on the elastic potential,

$$\mathcal{S}^{\text{elas}}[u] = \mathcal{P}[u] = \int_\Omega \frac{\mu}{4} \sum_{j,k=1}^{d} (\partial_{x_j} u_k + \partial_{x_k} u_j)^2 + \frac{\lambda}{2} (\operatorname{div} u)^2 \, dx;$$

$$\mathcal{A}^{\text{elas}}[u] = \mu \Delta u + (\lambda + \mu) \nabla \operatorname{div} u.$$

- Two parameters, the Lamé constants μ and λ describe material properties, (α may be incorporated), and typically $\lambda = 0$ (maximal expansion of the elastic body).
- Physically meaningful transformation.
- Transformation is restricted to small, local deformations.
- Fast $\mathcal{O}(N \log N)$ implementation based on FFT-type factorization for periodic boundary conditions.

Fluid registration

- Regularization is based on the elastic potential of the time derivative of the displacement, $\mathcal{S}^{\text{fluid}}[u] = \mathcal{P}[\partial_t u]$;

$$\mathcal{A}^{\text{elas}}[v] = f(\,\cdot\,, u), \quad v = \frac{d}{dt} u = \partial_t u + \nabla u \; v.$$

- Two parameters, the Lamé constants μ and λ describe material properties, (α may be incorporated), and typically $\lambda = 0$ (maximal expansion of the fluid).
- Additional Euler step.
- Physically meaningful transformation.
- Extremely flexible.
- Fast $\mathcal{O}(N \log N)$ implementation based on FFT-type factorization for periodic boundary conditions.

Diffusion registration

- Regularization is based on first order spatial derivatives of the displacement,

$$\mathcal{S}^{\text{diff}}[u] = \frac{1}{2} \sum_{\ell=1}^{d} \int_\Omega \langle \nabla u_\ell, \nabla u_\ell \rangle_{\mathbb{R}^d} \, dx;$$

$$\mathcal{A}^{\text{diff}}[u] = \Delta u.$$

- Two parameters, the regularizing parameter α and time-step τ.
- Transformation is restricted to small deformations.

- Can be combined with the fluid idea to allow for larger deformations.
- $\mathcal{O}(N)$ implementation based on an AOS scheme.

Curvature registration

- Regularization based on second order spatial derivatives of the displacement,

$$\mathcal{S}^{\text{curv}}[u] = \frac{1}{2}\sum_{\ell=1}^{d}\int_{\Omega}(\Delta u_\ell)^2\,dx;$$
$$\mathcal{A}^{\text{curv}}[u] = \Delta^2 u.$$

- Two parameters, the regularizing parameter α and time-step τ.
- Kernel contains affine linear transformations.
- Transformation is restricted to small deformations.
- Can be combined with the fluid idea to allow for larger deformations.
- Fast $\mathcal{O}(N\log N)$ implementation based on DCT-type techniques.

Note that the regularization of elastic, diffusion, and curvature registration can be applied to the displacement as well as to the update of the displacement. For elastic and fluid registration one may also use a time marching implementation whereas for diffusion and curvature registration one may also use a fixed-point iteration.

13.2 Timings for non-parametric registrations

The timings for two-dimensional image registrations of images of various sizes are summarized in Table 13.1, see also Fig. 13.1. To be precise, we present timings obtained for the solution of one linear system of equations for the respective registration schemes. Note that this is the time consuming part of the computations. The different implementations are by no means optimized. However, the implementations lead to a fair comparison of the different solution schemes.

For the linear systems of equations arising in the elastic registration we implemented four different solution schemes: a direct solver using MATLAB's \-operator (cf., Math Works (1992)), the conjugate gradient (CG) method of Hestenes Stiefel (1952), a V-cycle of a multigrid method (MG; cf., e.g., Hackbusch (1993, §10.4)), and the FFT-type solution scheme developed in Section 9.5. In our MG implementation, the coarsest grid is 1-by-1, and two pre-smoothing steps and one post-smoothing step with a Jacobi smoother and relaxation parameter $\theta = 2/3$ are performed.

As is apparent from Table 13.1 and Fig. 13.1, the direct solution scheme is not applicable for images of more than 64×64 pixels. The reason is that this direct solution schemes requires additional storage for the triangular factor of an LU decomposition of A; cf., eqn (9.24). Since A is a $2N$-by-$2N$ matrix with $N = n_1 n_2$, the U factor requires $16N^2$ Bytes storage, e.g., 4096 MBytes for

Table 13.1 Timings for the solution of one linear system of equations arising in non-parametric registrations. The timings are given in CPU seconds and are obtained on a PC with Intel Pentium III, 1000 MHz, 256 KBytes cache, and 256 MBytes RAM. The operating system used is SuSE LINUX 8.0 and the computations were performed under MATLAB 6.0.0.88 (R12); $*$ indicates out of memory or computation stopped after about 30 minutes; see also Fig. 13.1.

Image	Elastic registration				Diffusion registration			Curvature registration	
size	Direct	CG	MG	FFT	Direct	DCT	AOS	Direct	DCT
16^2	0.0710	0.0930	0.0510	0.0010	0.0060	0.0050	0.0030	0.0130	0.0060
32^2	0.8490	0.1210	0.0670	0.0020	0.0260	0.0020	0.0120	0.0740	0.0020
64^2	11.1880	0.2460	0.1050	0.0120	0.1570	0.0160	0.0490	0.6520	0.0110
128^2	$*$	1.0620	0.2770	0.0680	1.3430	0.0960	0.2240	6.4390	0.0920
256^2	$*$	4.4710	1.0760	0.3740	10.9040	0.6230	0.9360	72.2980	0.6240
512^2	$*$	17.1200	4.1980	1.5100	$*$	4.1290	3.7940	$*$	4.1850
1024^2	$*$	67.1450	16.8190	6.3980	$*$	29.2050	14.8570	$*$	28.7030

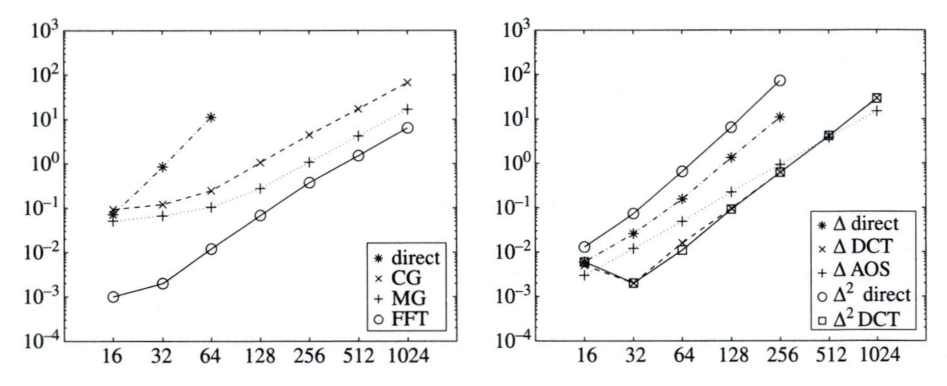

FIG. 13.1 Timings for the solution of one linear system of equations arising in non-parametric registrations. Timings are given in CPU seconds; LEFT: elastic registration with direct, CG, MG, and FFT solver, RIGHT: diffusion and curvature registration with direct, DCT-, and AOS-based solution schemes; see also Table 13.1.

$N = 128^2$. In addition, the direct solver does not work with Neumann or periodic boundary conditions, because of the singularity of the linear systems.

The CG iteration was stopped when the relative residual was brought below a user-supplied threshold ($\mathrm{tol}_{\mathrm{CG}} = 10^{-2}$), or the number of iterations increases beyond a user-supplied number ($k_{\max} = 100$). Unfortunately, in all our experiments the first criterion never applied. With this crude stopping rule, the complexity of CG is $\mathcal{O}(N)$; however, the constant is quite large.

For the linear MG, we applied only one V-cycle. The complexity of the linear MG is also $\mathcal{O}(N)$. However, the constant is better than the one for the CG iteration.

Finally, the FFT-based solution scheme has only complexity $\mathcal{O}(N \log N)$. Note that for small values of N (including all values presented in Table 13.1) this scheme is superior to the others. It is also worthwhile noticing that the FFT-based solution scheme produces exact solutions for the linear systems of equations, while the CG and MG schemes give only approximations. For the FFT-based scheme we have an additional initialization phase for the computation of the entries of the pseudo-inverse of $F^{\mathrm{H}}AF$; cf., Theorem 9.9.

For diffusion registration, we implemented a direct solver using MATLAB's \-operator (cf., Math Works (1992)), the DCT-based factorization (cf., Section 11.2), and an AOS-based solution scheme (cf., Section 11.3). Note that the system matrix for the diffusion registration is $I_d \otimes A^{\mathrm{diff},d}$ (cf., eqn (11.4)) and thus block diagonal. Moreover, from eqn (11.8) we have that $A^{\mathrm{diff},d}$ is a block tridiagonal matrix with a small bandwidth. However, the bandwidth grows with image size.

The direct solver is only applicable for small image sizes (up to 256^2) because of memory requirements. For the DCT-based solution scheme we have an initialization phase for the computation of $D^{(d)}$; cf., Theorem 11.3. For small values of N (up to $N = 512^2$) the $\mathcal{O}(N \log N)$ implementation performs faster than the AOS-based $\mathcal{O}(N)$ scheme.

Finally, we implemented a direct solver using MATLAB's \-operator (cf., Math Works (1992)), and a DCT-based factorization (cf., Section 12.1) for curvature registration. Since $A^{\mathrm{curv},d} = (A^{\mathrm{diff},d})^{\top} A^{\mathrm{diff},d}$, the bandwidth of the curvature matrix is larger than the bandwidth of the diffusion matrix. Thus, in accordance with our measurements, we expect that more time is required for the curvature than for the diffusion direct solution scheme. On the other hand, for the DCT-based solution scheme we expect exactly the same time for the diffusion and curvature registration schemes. For both registration schemes, a DCT, a multiplication with a diagonal matrix, and an inverse DCT have to be computed. The only difference is the entries of the diagonal matrices. Our numerical experiments almost reflect this expectation.

Summarizing, it turned out that for elastic and fluid registration the FFT-based direct solution scheme provides an attractive alternative to iterative solution schemes like CG and MG. In particular, for moderate image sizes, the FFT-based technique outperforms its competitors.

For diffusion registration, the DCT-based solution scheme is preferable for small image sizes while the AOS-based solution scheme is the fastest scheme for large-scaled images. The DCT-based solution scheme is at present the only competitive scheme for curvature registration.

Note that the FFT-based scheme is faster than the AOS-based solution scheme for all image sizes considered in Table 13.1. This is partly related to the excellent implementation of the FFT routine provided by Matlab (1992).

13.3 A competitive example: hands

In this section we present competitive results for elastic, fluid, diffusion, and curvature registration. We register the reference image and the linearly pre-registered template shown in Fig. 9.10.

For elastic registration, we chose $\mu = 500$ and $\lambda = 0$ and used the FFT-based solution scheme for a fixed-point-type implementation. For fluid registration we chose $\mu = 500$ and $\lambda = 0$, and used the FFT-based solution scheme for the velocity and the Euler step presented in Section 10.4.1. For diffusion registration we chose $\alpha = 5$ and $\tau = 0.25$, and for curvature registration we chose $\alpha = 50$ and $\tau = 0.5$. For diffusion registration we used a time marching AOS-based implementation and for curvature registration a time marching DCT-based implementation.

All registrations started with $u^{(0)} = 0$ and were stopped as soon as

$$\frac{\mathcal{D}^{\mathrm{SSD}}[R, T; u^{(k)}]}{\mathcal{D}^{\mathrm{SSD}}[R, T]} \le 0.6.$$

This stopping criterion was met for the values $k^{\mathrm{elas}} = 33, k^{\mathrm{fluid}} = 19, k^{\mathrm{diff}} = 12$, and $k^{\mathrm{curv}} = 11$. Note that these values depend on the choice of the parameters and therefore cannot be compared directly.

Figure 13.2 illustrates the displacements for the four techniques and Fig. 13.3 displays some details. As is apparent from these figures, all four registration schemes produced visually pleasing results. The differences between the four schemes are rather small in this example. In particular, the differences between elastic and fluid registration are hardly visible. As expected, diffusion registration yields a less smooth displacement field, because the regularizer is first order derivative based. Note that the results of the registration schemes may differ considerably for other examples; see, e.g., the • to C registration. However, if the images are very similar, i.e., show only minor differences, the results of the registration schemes are also very similar.

13.4 How to compare apples with pears

It has already been pointed out that image registration is an ill-posed problem. In particular, in Section 12.3 we presented two different registrations of two images of a square. Since the difference between the reference and the deformed template is zero for both registrations, both approaches are reasonable and equally likely. Without additional, external information, we are not able to decide which of the two registrations is to be preferred. In fact both are equally wrong, because the template image is a rotated copy of the reference (see also the discussion in Section 12.3).

Thus, without additional, external knowledge, registration is a risky task. At this point regularization enters into the picture. Regularization not only is a mathematical technique to overcome ill-posedness of a problem, but in

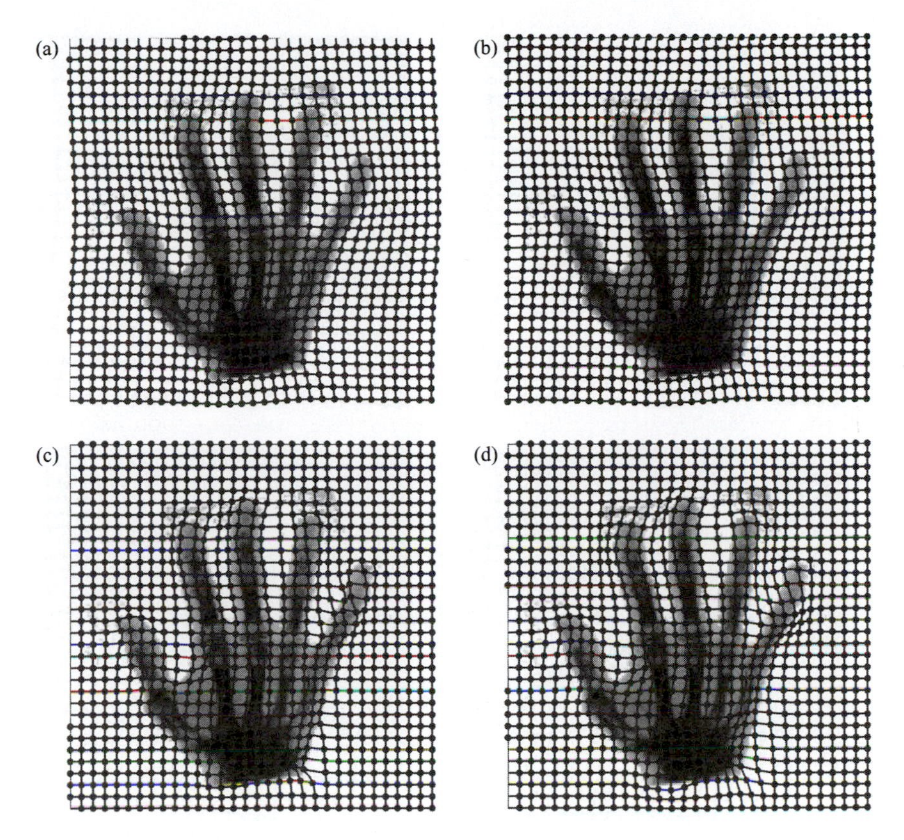

FIG. 13.2 Results of elastic (a), fluid (b), diffusion (c), and curvature (d) registration.

image registration, it also presents a way of providing external information. For example, if we know that the distortions of an image are mostly due to elastic deformations, we should use a regularizer based on the elastic potential. However, if we only have a vague idea about the source of the observed distortions, the elastic registration scheme may turn out to be the wrong choice. In this situation, a straightforward idea would be to apply the most flexible scheme in the hope of reducing the distance measure at most. Note that this flexibility enables one to register very different images; cf., e.g., Section 10.7. Though this truly is a very attractive option for some applications, this capability is frightening for others. For example, it is easy to obtain visually pleasing results for the registration of two different human brains even when one of them has been rotated by 90 or 180 degrees by accident.

Though the differences between the different regularizers might be large for some registrations (i.e., the • to C registration), they are in general small for

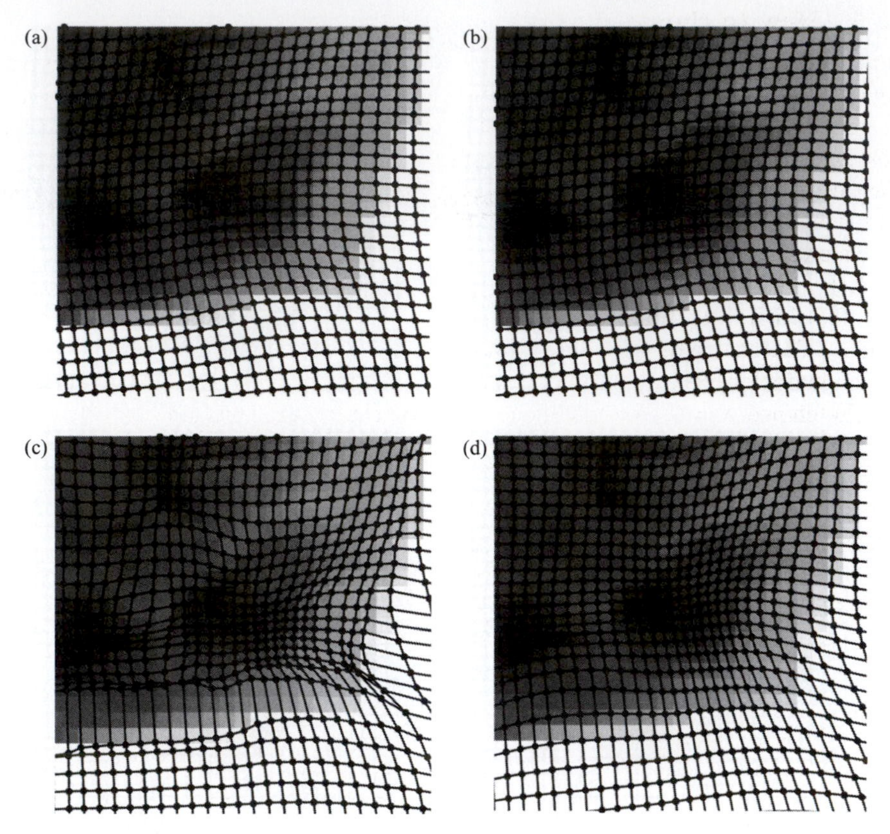

FIG. 13.3 Results of elastic (a), fluid (b), diffusion (c), and curvature (d) registration, details.

closely related images; cf., Section 13.3. Thus, one may argue that it is just a matter of taste which regularizer is to be used. However, depending on the application even small differences may be of importance and one always should use the best available model.

Another important question is, which distance measure should be used? Again the answer depends strongly on the particular application. Here, using a particular and appropriate distance measure can also help to reduce unwanted local minima; see also the discussion in Section 6.6. Moreover, one may also use the distance measure as an important modeling tool in order to provide additional, external knowledge to the registration procedure.

The solution to all these questions is probably to provide a variety of different registration techniques and to enable one to choose the most promising technique for a particular application.

13.5 How to choose μ and λ or α and τ

How does one choose the parameters in non-parametric image registration? Again this is a delicate question. Note that in image registration the focus is in general not on a perfect physical model but on a meaningful deformation. For example, in the HNSP (cf., Chapter 2) one has to deal not only with pure elastic deformations of the paraffin wax embedded tissue but also with therm and dehydration processes. Hence, solving the Navier–Lamé equations to high precision is probably not the appropriate way.

Let us start with elastic and fluid registration. In our experiments, we chose $\lambda = 0$, which is a common choice; cf., e.g., Broit (1981) or Bajcsy & Kovačič (1986). The choice of μ in this book is always explicitly stated. However, in our implementation we generally use an automatic procedure to choose μ. Note that for the choice $\lambda = 0$, the matrix stencils for elastic registration (cf., Table 9.1) are linear with respect to μ. Since the scaling of the force field is in general unknown, there is always one additional degree of freedom in our system. For our particular implementation, the forces are computed using a central finite difference approximation, i.e.,

$$F_i^{k,\ell} = \gamma_\ell \Big(R_i - T_i^{(k)}\Big)\Big(T_{i+e_\ell}^{(k)} - T_{i-e_\ell}^{(k)}\Big), \quad 1 \le \ell \le d,$$

where $i = (i_1, \ldots, i_d) \in \mathbb{N}^d, 1 \le i_p \le n_p, 1 \le p \le d$, and e_ℓ denotes the ℓth unit vector in $\mathbb{R}^d$. For a central finite difference, the standard choice for the scaling factor is $\gamma_\ell = (2h_\ell)^{-1}$, where $h_\ell = n_\ell^{-1}$ is the meshsize. In our implementation, we use $\gamma_\ell = n_\ell/(2 \cdot 255^2)$ which implicitly re-scales the gray value range from $[0, 255]$ to $[0, 1]$. Note that the displacement is computed from the force field via a linear system of equations. Thus, we are able to provide a heuristic for a clever choice of μ which we briefly describe.

To estimate μ, we define a tolerance for the maximal displacement of a pixel (we use $u_{\max} = 2$ in our computations). In the first step, we set $\mu = 1$ and $\lambda = 0$. We then compute the forces, solve the linear system of equations for a displacement $\hat{u}$, and compute the norm of this displacement. The estimator for μ is simply $\mu = u_{\max}/\left\|\hat{u}\right\|_V$ (cf., eqn (10.7)). We re-scale the matrix A^{elas} and set $u^{(1)} = \mu\hat{u}$. With this choice and the re-scaling of the displacement, we have $\left\|u^{(1)}\right\|_V = u_{\max}$. We take this value of μ during the computations but monitor the norm of the displacements. If $\left\|u^{(k)}\right\|_V > u_{\max}$, we re-scale the actual displacement such that the norm of the re-scaled displacement becomes $u_{\max}$. However, the value of μ is not altered. In our experiments (not contained in this book), this procedure works surprisingly well.

In principle the same remarks and ideas apply also to the time marching implementation for diffusion and curvature registration. Here, roughly speaking, the linear systems

$$(I - \tau\alpha A)\vec{U}^{(k+1)} = \vec{U}^{(k)} + \tau\vec{F}^{(k)}$$

are to be solved. Note that on the right hand side τ compromises between the current displacements and the current forces. Thus, a re-scaling of the forces also affects the choice of τ. Unfortunately, the matrix $I - \tau\alpha A$ is not simply linear in α or τ. Thus, a direct application of the above strategy cannot be recommended.

Another interesting strategy is to relax the regularizer during iteration; see also Henn (1997). For example, for the time marching implementation, we may use $\alpha = \alpha(t)$ where $\alpha(t) \to 0$ for $t \to \infty$. The motivation is that the regularizer should guide the registration to the required minimizer of the distance measure. Getting closer to this minimizer reduces the impact of the regularizer. Of course, it is also possible to take $\alpha = \alpha(x,t)$, which has in principle already been discussed in Section 11.6.2 for the demons registration. Note that for this choice, the matrices arising from the discrete formulation cannot be factorized explicitly.

13.6 Further accelerations of the schemes

Multi-resolutions and parallelizing techniques present further possibilities for a reduction of computation time. The first step of a multi-resolution procedure is to provide a pyramid of images with decreasing resolution. Starting with the coarsest resolution, one computes the transformation with relatively low computational costs. The coarse resolution transformation is prolongated onto the next finer grid and serves as a typically excellent starting point for the registration at the finer resolution. Because of this good starting point, one expects lower computational costs for corrections on the finer grid. This procedure is continued until the finest resolution is reached. Note that this is also the basic idea behind multigrid techniques; cf., e.g., Hackbusch (1993, §10.4), Henn (1997), or Schormann et al (1996).

For the three-dimensional reconstruction of the human brain presented in Section 9.10, we used a Gaussian pyramid approach. With this approach the computational time for a two-dimensional registration of two 1024^2 pixel images could be reduced from about 20 minutes to about 2 minutes.

Furthermore, a multi-resolution approach not only reduces the computational costs but also serves as an additional regularizer. The resolution hierarchy also introduces a hierarchy for the images differences. Major differences are visible in all resolutions while minor differences appear only at finer scales. In this sense, the registration process is guided. Essential differences are removed on the corse grid which also prevents the scheme stagnating in local minima.

Parallelizing techniques provide a further possibility for accelerating the computations. Unfortunately, these techniques have to be adapted to the particular computational environment as well as to the particular schemes. Thus, a description is in general very technical and therefore we refer to the literature: for elastic registration see, e.g., Christensen et al (1996) or Modersitzki et al (2001), for FFT-based elastic registration see Böhme et al (2002), and for AOS-based diffusion registration see Bolten et al (2003).

It should be noted that also higher order approximations for the non-linear forces (cf., eqn (8.11)) can be considered. But, as has already been pointed out in Chapter 6, the numerical computation of higher order derivatives can be critical in image processing.

13.7 Extensions

The proposed unified approach to image registration can be extended in various directions. For example, a symmetrized distance measure

$$\mathcal{D}^{\text{sym}}[R,T;\varphi] = \mathcal{D}[R,T;\varphi] + \mathcal{D}[T,R;\varphi^{-1}]$$

(see, e.g., Christensen & Johnson (2001)) may be treated analogously. This can be of interest if the template image is not a deformed version of an ideal (as, e.g., in the HNSP; cf., Chapter 2) but reference and template images are on a par. Moreover, one may also introduce ψ as an approximation to φ^{-1}. A variational formulation might be based on the functional

$$\mathcal{D}[R,T;\varphi] + \mathcal{D}[T,R;\psi] + \alpha\mathcal{S}[\varphi,\psi],$$

where $\mathcal{S}[\varphi,\psi]$ penalizes the deviation between ψ and φ^{-1}, e.g.,

$$\mathcal{S}[\varphi,\psi] = \int_\Omega \Big(1 - \det(\varphi\circ\psi(x))\Big)^2 dx,$$

or

$$\mathcal{S}[\varphi,\psi] = \int_\Omega \|\varphi(x) - \psi^{-1}(x)\|^2_{\mathbb{R}^d} + \|\varphi^{-1}(x) - \psi(x)\|^2_{\mathbb{R}^d}\, dx.$$

For other applications, like, for example, the registration of MRI scans from different phases of a contrast agent uptake, it can be of particular importance to maintain the volume of the tissue. This goal can be achieved, for example, by using the additional regularizer

$$\mathcal{S}^{\text{vol}}[\varphi] = \int_\Omega \det(\nabla u)\, dx.$$

Note that $\det(\nabla\varphi) = 1 + \det(\nabla u)$ and the transformation maintains the volume, if $\det(\nabla\varphi) = 1$.

Recent research is also concerned with the integration of interpolation constraints to non-parametric image registration. Here, the idea is to supply a number of landmarks which are of particular interest for the underlying application. The required transformation should minimize the joint functional subject to some interpolation constraints or extended by an additional penalizing term which particularly enforces a match of the landmarks.

REFERENCES

Abram, K. A. (2000). Registration of images with dissimilar contrast using a hybrid method employing correlation and mutual information. Technical report, Dept. of Computer Science, Dartmouth College, Hannover.

Aldroubi, A. and Gröchening, K. (2001). Nonuniform sampling and reconstruction in shift-invariant spaces. *SIAM Review 43*(3), 585–620.

Alpert, N. M., Bradshaw, J. F., Kennedy, D., and Correia, J. A. (1990). The principal axes transformation – A method for image registration. *Journal of Nuclear Medicine 31*(10), 1717–1722.

Amit, Y. (1994). A nonlinear variational problem for image matching. *SIAM Journal of Scientific Computing 15*(1), 207–224.

Bajcsy, R. and Kovačič, S. (1986). Toward an individualized brain atlas elastic matching. Technical Report MS-CIS-86-71 Grasp Lap 76, Dept. of Computer and Information Science, Moore School, University of Philadelphia.

Böhme, M., Hagenau, R., Modersitzki, J., and Siebert, B. (2002). Non-linear image registration on PC-clusters using parallel FFT techniques. Preprint, Institute of Mathematics, University of Lübeck, Germany.

Bolten, M., Modersitzki, J., and Papenberg, N. (2003). A parallel implementation of the AOS-based diffusion registration. Preprint, in preparation, Institute of Mathematics, University of Lübeck.

Boor, C. De (1978). *A practicle guide to splines.* Springer, New York.

Braess, D. (1980). *Nonlinear approximation theory,* Volume 7 of *Springer Series in Computational Mathematics.* Springer.

Brewer, J. W. (1978). Kronecker products and matrix calculus in system theory. *IEEE Transactions on Circuits and Systems 25*, 772–780.

Broit, C. (1981). *Optimal Registration of Deformed Images.* Ph.D. thesis, Computer and Information Science, University of Pensylvania.

Bro-Nielsen, M. (1996). *Medical Image Registration and Surgery Simulation.* Ph.D. thesis, IMM, Technical University of Denmark.

Bro-Nielsen, M. and Gramkow, C. (1996). Fast fluid registration of medical images. *Lecture Notes in Computer Science*, *1131*, 267–276, Springer, Berlin and Heidelberg.

Brown, L. G. (1992). A survey of image registration techniques. *ACM Computing Surveys 24*(4), 325–376.

Buck, R. C. (1978). *Advanced Calculus.* McGraw-Hill, New York.

Čapek, M. (1999). Optimisation strategies applied to global similarity based image registration methods. In *WSCG '99: the 7th International Conference in Central Europe on Computer Graphic*, pp. 369–374.

Christensen, G. E. (1994). *Deformable Shape Models for Anatomy.* Ph.D. thesis, Sever Institute of Technology, Washington University.

Christensen, G. E. and Johnson, H. J. (2001). Consistent image registration. *IEEE Transactions on Medical Imaging 20*(7), 568–582.

Christensen, G. E., Miller, M. I., Grenander, U., and Vannier, M. W. (1996). Individualizing neuroanatomical atlases using a massively parallel computer. *IEEE Computer*, 32–38.

Collignon, A., Vandermeulen, A., Suetens, P., and Marchal, G. (1995). 3d multi-modality medical image registration based on information theory. *Computational Imaging and Vision 3*, 263–274, Kluwer Academic, Dordrechth.

Collins, D. L. and Evans, A. C. (1997). Animal: Validation and applications of nonlinear registration-based segmentation. *International Journal of Pattern Recognition and Artificial Intelligence 11*(8), 1271–1294.

Curtis, P. C. (1958). n-parameter families and best approximation. *Pacific Journal of Mathematics 58*, 1013–1027.

Davis, P. J. (1979). *Circulant Matrices.* Chelsea Publishing, New York.

Duchon, J. (1976). Interpolation des fonctions de deux variables suivant le principle de la flexion des plaques minces. *RAIRO Analyse Numérique 10*, 5–12.

Duda, R. and Hart, P. (1973). *Pattern Classification and Scene Analysis.* Wiley.

Dümbgen, L. (2001). Private communication.

Fischer, B. and Modersitzki, J. (1999). Fast inversion of matrices arising in image processing. *Numerical Algorithms 22*, 1–11.

Fischer, B. and Modersitzki, J. (2003). Curvature based image registration. *Journal of Mathematical Imaging and Vision 18*(1), 81–85.

Fletcher, R. (1987). *Practical Methods of Optimization.* Wiley.

Folland, G. B. (1995). *Introduction to partial differential equations* (2nd edn). Princeton University Press, Princeton, New Jersey.

Gaens, T., Maes, F., Vandermeulen, D., and Suetens, P. (1998). Non-rigid multimodal image registration using mutual information. In *Medical Image Computing and Computer-Assisted Intervention – MICCAI '98* (ed. W. M. Wells, A. Colchester, and S. Delp), pp. 1099–1106. Springer.

Golub, G. H. and van Loan, C. F. (1989). *Matrix Computations* (2nd edn). The Johns Hopkins University Press, Baltimore, Maryland.

Gonzalez, R. C. and Woods, R. E. (1993). *Digital Image Processing.* Addison-Wesley, Reading, Massachusetts.

Gurtin, M. E. (1981). *An Introduction to Continuum Mechanics.* Academic Press, Orlando, Florida.

Gurtin, M. E. (1983). *Topics in Finite Elasticity.* SIAM, Philadelphia, Pennsylvania.

Hackbusch, W. (1987). *Partial Differential Equations.* Teubner, Stuttgart.

Hackbusch, W. (1993). *Iterative Solution of Large Sparse Systems of Equations.* Springer, New York.

Henn, S. (1997). *Schnelle elastische Anpassung in der digitalen Bildverarbeitung mit Hilfe von Mehrgitterverfahren.* Ph.D. thesis, Heinrich-Heine-Universität, Düsseldorf.

Hestenes, M. R. and Stiefel, E. (1952). Methods of conjugate gradients for solving linear systems. *Journal of Research of the National Bureau of Standards 49*, 409–436.

Horn, B. K. P. and Schunck, B. G. (1981). Determining optical flow. *Artificial Intelligence 17*, 185–204.

Horn, R. A. and Johnson, C. R. (1990). *Matrix Analysis.* Cambridge University Press, Cambridge.

Horn, Roger A. and Johnson, Charles R. (1991). *Topics in Matrix Analysis.* Cambridge University Press, Cambridge.

Keeling, S. L. and Ring, W. (2002). Medical image registration and interpolation by optical flow with maximal rigidity. SFB Report No. 248, Insitute of Mathematics, Karl-Franzens-Universität Graz.

Kent, J. T. and Tyler, D. E. (1988). Maximum likelihood estimation for the wrapped cauchy distribution. *Journal of Applied Statistics 15*(2), 247–254.

Kent, J. T. and Tyler, D. E. (1991). Redescending m-estimates of multivariate location and scatter. *Annals of Statistics 19*(4), 2102–2119.

Kim, B., Boes, J. L., Frey, K. A., and Meyer, C. R. (1997). Mutual information for automated unwarping of rat brain autoradiographs. *Neuroimage 5*, 31–40. Article No. N1960251.

Kullback, S. and Leibler, R. A. (1951). On information and sufficiency. *Annals of Mathematical Statistics 22*, 79–86.

Light, W. A. (1995). Variational methods for interpolation, particularly by radial basis functions. *Numerical Analysis*, 94–106.

Maes, F., Collignon, A., Vandermeulen, D., Marchal, G., and Suetens, P. (1997). Multimodality image registration by maximization of mutual information. *IEEE Transactions on Medical Imaging 16*(2), 187–198.

Maintz, J. B. A. and Viergever, M. A. (1998). A survey of medical image registration. *Medical Image Analysis 2*(1), 1–36.

Mairhuber, J. C. (1956). On Haar's theorem concerning Chebychev approximation problems having unique solutions. *Proceedings of the AMS 7*, 609–615.

Mardia, K., Kent, J., and Bibby, J. (1979). *Multiavariant Analysis.* Academic Press, New York.

MathWorks (1992). *MATLAB User's Guide.* MathWorks, Natick, Massachusetts.

Maurer, C. R. and Fitzpatrick, J. M. (1993). A review of medical image registration. In *Interactive Image-Guided Neurosurgery*, pp. 17–44. American Association of Neurological Surgeons, Park Ridge.

Meihe, X., Srinivasan, R., and Nowinski, W. L. (1999). A fast mutual information method for multi-modal registration. In *Information processing in medical imaging* (ed. A. Kuba, M. Šámal, and A. Todd-Pokropek), pp. 466–471. Springer Berlin and Heidelberg.

Modersitzki, J., Lustig, G., Schmitt, O., and Obelöer, W. (2001). Elastic registration of brain images on large PC-clusters. *Future Generation Computer Systems 18*, 115–125, Elsevier, Amsterdam.

Nocedal, J. and Wright, S. J. (1999). *Numerical optimization.* Springer, New York.

Pennec, X., Cachier, P., and Ayache, N. (1999). Understanding the "demon's algorithm" 3D non-rigid registration by gradient descent. In *Medical image computing and computer assisted intervention* (ed. C. Taylor and A. Colchester), pp. 597–605. Springer, Berlin and Heidelberg.

Piegl, L. and Tiller, W. (1997). *The* NURBS *book* (2nd edn). Springer.

Roche, A., Malandain, G., Ayache, N., and Prima, S. (1999). Towards a better comprehension of similarity measures used in medical image registration. In *Medical image computing and computer assisted intervention* (ed. C. Taylor and A. Colchester), pp. 555–566. Springer, Berlin and Heidelberg.

Rohr, K. (2001). Landmark-based image analysis. *Computational Imaging and Vision.* Kluwer Academic, Dordrecht.

Schaback, R. (1997). Reconstruction of multivariate functions from scattered data. `http://www.num.math.uni-goettingen.de/schaback/teaching.html`.

Schmitt, O. (2001). *Die multimodale Architektonik des menschlichen Gehirns.* Habilitation, Insitute of Anatomy, Medical University of Lübeck.

Schormann, T., Henn, S., and Zilles, K. (1996). A new approach to fast elastic alignment with applications to human brains. *Lecture Notes in Computer Science 1131*, 337–342, Berlin and Heidelberg.

Schormann, T. and Zilles, K. (1997). Limitations of the principal-axes theory. *IEEE Transactions on Medical Imaging 16*(6), 942–947.

Shannon, C. E. (1949). Communications in the present of noise. *Proceedings of the IRE 37*, 10–21.

Studholme, C., Hill, D. L. G., and Hawkes, D. J. (1996). Automated 3-D registration of MR and CT images of the head. *Medical Image Analysis 1*(2), 163–175.

Thévenaz, P., Blu, T., and Unser, M. (2000). Image interpolation and resampling. In *Handbook of Medical Imaging, Processing and Analysis* (ed. I. Bankman), pp. 393–420. Academic Press, San Diego, California.

Thévenaz, P., Ruttimann, U. E., and Unser, M. (1998). A pyramid approach to subpixel registration based on intensity. *IEEE Transactions on Image Processing 7*(1), 27–41.

Thirion, J.-P. (1995). Fast non-rigid matching of 3D medical images. Technical Report 2547, Institut National de Recherche en Informatique et en Automatique, France.

Thirion, J.-P. (1998). Image matching as a diffusion process: An analogy with Maxwell's demons. *Medical Image Analysis 2*(3), 243–260.

Thomas, L. H. (1949). Elliptic problems in linear difference equations over a network. Technical report, Watson Scientific Computing Laboratory, Columbia University, New York.

van den Elsen, P. A., Pol, E.-J. D., and Viergever, M. A. (1993). Medical image matching – a review with classification. *IEEE Engineering in Medicine and Biology*, 26–38.

Viola, P. A. (1995). *Alignment by Maximization of Mutual Information.* Ph.D. thesis, Massachusetts Institute of Technology.

Weickert, J. (1998). *Anisotropic diffusion in image processing.* Teubner, Stuttgart.

Wells III, W. M., Viola, P., Atsumi, H., Nakajima, S., and Kikinis, R. (1996). Multi-modal volume registration by maximization of mutual information. *Medical Image Analysis 1*(1), 35–51.

Wollny, G. and Kruggel, F. (2002). Computational cost of non rigid registration algorithms based on fluid dynamics. *IEEE Transactions on Medical Imaging, 21*(8), 946–952.

GLOSSARY

Norms and such

$[x]$	$[x] := \min\{j \in \mathbb{Z} \mid j - 0.5 \le x < j + 0.5\}$.
$\lceil x \rceil$	$\lceil x \rceil := \min\{j \in \mathbb{Z} \mid j \ge x\}$.
$\lfloor x \rfloor$	$\lfloor x \rfloor := \max\{j \in \mathbb{Z} \mid j \le x\}$.
$\|\cdot\|_2$	For $A \in \mathbb{C}^{n\times n}$, $\|A\|_2 := \max\{\sqrt{\lambda} : \lambda \text{ is eigenvalue of } A^{\mathrm{H}}A\}$.
$\|\cdot\|_f$	norm on a feature space.
$\|\cdot\|_{\mathbb{R}^d}$	norm on $\mathbb{R}^d$, $\|x\|_{\mathbb{R}^d} := \sqrt{\langle x, x\rangle_{\mathbb{R}^d}}$.
$\langle\cdot,\cdot\rangle_{\mathbb{R}^d}$	inner product on $\mathbb{R}^d$, $\langle x, y\rangle_{\mathbb{R}^d} := y^\top x$.
$\langle\cdot,\cdot\rangle_0$	$\langle f, g\rangle_0 := \int_{\mathbb{R}^d} f(x)g(x)\,dx$.
$\langle\cdot,\cdot\rangle_{L_2}$	$\langle f, g\rangle_{L_2} := \int_{\mathbb{R}^d} f(x)g(x)\,dx$.
$\langle\cdot,\cdot\rangle_q$	semi-inner product, $\langle f, g\rangle_q := \sum_{\lvert\kappa\rvert=q} c_\kappa \langle D^\kappa f, D^\kappa g\rangle_0$.

Attributes

$\dagger$	$B^\dagger$ Moore–Penrose pseudo-inverse.
H	Hermitian, for $A = (a_{j,k})_{j,k} \in \mathbb{C}^{n\times n}$, $A^{\mathsf{H}} = (\overline{a_{k,j}})_{j,k}$.
$\perp$	$x \perp y \iff \langle x, y\rangle_{\mathbb{R}^d} = 0$.
$\top$	transpose, for $A = (a_{j,k})_{j,k} \in \mathbb{R}^{n\times n}$, $A^\top = (a_{k,j})_{j,k}$.
$\vec{\ }$	$\vec{X} \in \mathbb{R}^N$, recollection of $X \in \mathbb{R}^{n_1\times\cdots\times n_d}$ in lexicographical ordering, $N = n_1\cdots n_d$.

Derivatives

∇^2	Hessian matrix $\nabla^2 f = (\partial_{x_j,x_k} f)_{j,k} \in \mathbb{R}^{d\times d}$.
∇	$\nabla f = (\partial_{x_1} f, \ldots, \partial_{x_d} f)^\top$ for functions $f : \mathbb{R}^d \to \mathbb{R}$, $\nabla\varphi = (\partial_{x_k}\varphi_j)_{j,k} \in \mathbb{R}^{d\times d}$ for vector fields $\varphi : \mathbb{R}^d \to \mathbb{R}^d$.
∂_{x_j}	partial derivative with respect to x_j.

Operators

$*$	convolution.
$\otimes$	Kronecker product.
$\times$	Cartesian product of spaces, also cross-product in $\mathbb{R}^3$.

Sets

$\partial\Omega$	boundary of $\Omega =]0,1[^d$.

A

AOS	additive operator splitting.

C

$\mathbb{C}$	complex numbers.
c_B	center of an image, $c_B := \mathbb{E}_B[x] \in \mathbb{R}^d$.
CCD	charge-coupled device.
C_m	cosine matrix, $C_m := \left(\cos \frac{(2j+1)k\pi}{2m}\right)_{j,k=0,\ldots,m-1} \in \mathbb{R}^{m\times m}$, also: basic circulant matrix, $C_m \in \mathbb{R}^{m\times m}$.
Corr	correlation, $\mathrm{Corr}_{R,T}(y) := \int_{\mathbb{R}^d} R(x)T(x-y)\,dx$.
$\mathbf{Cov}_B$	covariance, $\mathrm{Cov}_B := \mathbb{E}_B\left[(x-c_B)(x-c_B)^\top\right] \in \mathbb{R}^{d\times d}$.

D

$d\mathcal{F}[\varphi;\psi]$	Gâteaux derivative.
d	spatial dimension.
$\mathcal{D}^{\mathrm{corr}}$	correlation, $\mathcal{D}^{\mathrm{corr}}[R,T] := \left\langle \frac{R-\mu(R)}{\sigma(R)}, \frac{T-\mu(T)}{\sigma(T)} \right\rangle_{L_2}$.
DCT	discrete cosine transformation.
d/dt	total time derivative.
Δ	Laplace operator, $\Delta f = \sum_{j=1}^{d} \partial_{x_j,x_j} f$, $\Delta u = (\Delta u_1, \ldots, \Delta u_d)^\top$ for a vector field u.
δ_x	point evaluation functional at x, $\delta_x[f] = \int_{\mathbb{R}^d} f(y)\delta_x(y)\,dx = f(x)$.
det	determinant.
div	divergence, $\operatorname{div} v = \langle \nabla, v\rangle_{\mathbb{R}^d} = \sum_{j=1}^d \partial_{x_j} v_j$.
D^κ	$D^\kappa f := \left(\frac{\partial}{\partial x_1}\right)^{\kappa_1} \cdots \left(\frac{\partial}{\partial x_d}\right)^{\kappa_d} f$, $\kappa \in \mathbb{N}_0^d$.
$\mathcal{D}^{\mathrm{LM}}[\varphi]$	landmark-based distance measure, $\mathcal{D}^{\mathrm{LM}}[\varphi] := \sum_{j=1}^{m} \lVert \mathcal{F}(R,j) - \varphi(\mathcal{F}(T,j)) \rVert_f^2$.
D_m	$D_m := 2\,\mathrm{diag}(\cos\frac{k\pi}{m},\ k = 0,\ldots,m-1) \in \mathbb{R}^{m\times m}$.
$\mathcal{D}^{\mathrm{MI}}$	mutual information, $\mathcal{D}^{\mathrm{MI}}[R,T] = H(\rho_R) + H(\rho_T) - H(\rho_{R,T})$.
d_q	$d_q := \dim(\Pi_{q-1}(\mathbb{R}^d))$.
$\mathcal{D}^{\mathrm{SSD}}$	sum of squared differences, $\mathcal{D}^{\mathrm{SSD}}[R,T] := \frac{1}{2}\lVert T-R\rVert_{L_2}^2 = \frac{1}{2}\int_{\mathbb{R}^d} \left(T(x)-R(x)\right)^2\,dx$.

E

e_j	j^{th} column of the identify matrix I_d.
$\mathbb{E}_B[\,\cdot\,]$	expectation value, $\mathbb{E}_B[f] := \frac{\int_{\mathbb{R}^d} f(x)\,B(x)dx}{\int_{\mathbb{R}^d} B(x)dx}$.

F

FBS	flat bed scan.
FFT	fast Fourier transformation.
F_n	Fourier matrix, $F_n := n^{-1/2}(\omega_n^{(j-1)(k-1)})_{j,k=1,\dots,n} \in \mathbb{C}^{n\times n}$.
$F(S,j)$	j^{th} feature of an image S.

G

$\gamma(R,T;y)$	correlation coefficient, $\gamma(R,T;y) := \left\langle \frac{R-\mu(R)}{\sigma(R)}, \frac{T_y-\mu(T_y)}{\sigma(T_y)} \right\rangle_{L_2}$.

H

HNSP	Human NeuroScanning Project.
$H(\rho)$	entropy, $H(\rho) := -\mathbb{E}_\rho[\log\rho] = -\int_{\mathbb{R}^q} \rho\log\rho\,dg$.

I

I_d	identity matrix in $\mathbb{R}^{d\times d}$.
i.i.d.	independently and identically distributed.
Img(d)	set of images, $\mathrm{Img}(d) := \{\, b : \mathbb{R}^d \to \mathbb{R} \mid b \text{ is } d\text{-dimensional image} \,\}$.

L

$L_2(\Omega)$	$L_2(\Omega) := \{\, f : \Omega \to \mathbb{R} : \int_\Omega \lvert f(x)\rvert^2\,dx < \infty \,\}$.

M

MRI	magnetic resonance imaging.
$\mu(B)$	mean gray value, $\mu(B) := \lvert\Omega\rvert^{-1}\int_\Omega B(x)dx$.

N

$\mathbb{N}$	$\mathbb{N} = \{\,1,2,\dots\,\}$.
N	$N = n_1\cdots n_d$, number of grid points.
$\vec{n}$	outer normal vector on a boundary, $\vec{n} = (n_1,\dots,n_d)^\top$, $\lVert\vec{n}\rVert_{\mathbb{R}^d} = 1$.
$\mathbb{N}_0$	$\mathbb{N}_0 = \{\,0,1,2,\dots\,\}$.

O

$\mathcal{O}$	shortcut for a function belonging to a class of certain decay, $f : \mathbb{R}\to\mathbb{R}$ is $\mathcal{O}(h^p)$ if $\lim_{h\to 0} f(h)h^{-p-1} = 0$.
Ω	$\Omega =]0,1[^d$.
ω_n	n^{th} root of unity, $\omega_n := e^{-2\pi i/n}$.
Ω_n	$\Omega_n := \mathrm{diag}(\omega_n^0,\dots,\omega_n^{n-1})$.

P

PAT principal axes transformation.

φ transformation $\varphi : \mathbb{R}^d \to \mathbb{R}^d$.

$\Pi_q^d(\mathbb{R}^d)$ d-variate d-dimensional polynomials of degree q
$\Pi_q^d(\mathbb{R}^d) := \{ \varphi : \mathbb{R}^d \to \mathbb{R}^d \mid \varphi_\ell \in \Pi_q(\mathbb{R}^d),\ \ell = 1, \dots, d \}$.

$\Pi_q(\mathbb{R}^d)$ d-dimensional polynomials of degree q
$\Pi_q(\mathbb{R}^d) := \left\{ \psi : \mathbb{R}^d \to \mathbb{R} \mid \psi(x) = \sum_{|\kappa| \le q} \alpha_\kappa x^\kappa,\ \alpha_\kappa \in \mathbb{R} \right\}$.

ppi parts per inch.

$\mathcal{P}[u]$ elastic potential,

$$\mathcal{P}[u] = \int_\Omega \frac{\mu}{4} \sum_{j,k=1}^{d} (\partial_{x_j} u_k + \partial_{x_k} u_j)^2 + \frac{\lambda}{2} (\operatorname{div} u)^2 dx,$$

λ, μ Lamé constants.

R

R reference image.

$\mathbb{R}$ real numbers.

S

$\sigma(B)$ standard deviation, $\sigma(B) := \mu\big((B - \mu(B))^2\big)$.

$\Sigma(x,t)$ stress tensor, $\Sigma(x,t) := (\sigma_{j,k}(x,t))_{j,k}$.

SSD sum of squared differences.

$\mathcal{S}^{\mathrm{TPS}}$ $\mathcal{S}^{\mathrm{TPS}}[\psi] := \frac{1}{2} \langle \psi, \psi \rangle_q^2$.

T

T template image.

T_φ deformed template $T_\varphi(x) = T \circ \varphi(x) = T\big(\varphi(x)\big)$.

TPS thin plate spline.

trace trace of a matrix $A \in \mathbb{R}^{n \times n}$, $\operatorname{trace}(A) = \sum_{j=1}^n a_{j,j}$.

T_u deformed template $T_u(x) = T(x - u(x))$.

U

u displacement, $u(x) = \varphi(x) - x$.

V

v velocity, $v(P,t) = \partial_t u(P,t)$.

V_m $V_m := C_m \operatorname{diag}(\sqrt{1/m}, \sqrt{2/m}, \dots, \sqrt{2/m}) \in \mathbb{R}^{m \times m}$.

$V(x,t)$ strain tensor, $V(x,t) := (\varepsilon_{j,k}(x,t))_{j,k}$.

AUTHOR INDEX

SUBJECT INDEX